Steve Ford, WB8IMY, Editor

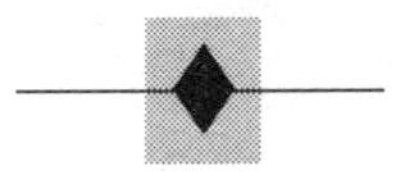

Published by: **The American Radio Relay League**
225 Main Street, Newington, CT 06111

Cover Photos: Kirk Kleinschmidt, NTØZ

Printed in USA

First Edition
First Printing

$8.00 in USA

ISBN: 0-87259-387-8

Foreword

Your VHF Companion introduces you to the wonders of VHF in an easy to understand, entertaining format. That doesn't mean we've skimped on the details, however. You'll find plenty of information, enough to get you started on the road to becoming an accomplished VHF operator.

If you're a new ham, this is the perfect book to have at your side as you explore your VHF privileges. If you've been a VHF operator for some time, *Your VHF Companion* may entice you to try something different. Regardless of your perspective, I hope the theme of this book will be clear: VHF is *fun*! Get on and enjoy the bands. You'll be surprised at how far VHF will take you!

David Sumner, K1ZZ
Executive Vice President
May 1992

Acknowledgments

Your VHF Companion would not be complete without recognizing the valuable behind-the-scenes contributions of several individuals: Jon Bloom, KE3Z, Bill Brown, WB8ELK, Luck Hurder, KY1T, Joel Kleinman, N1BKE, Joseph Moell, KØOV and Bob Schetgen, KU7G.

Contents

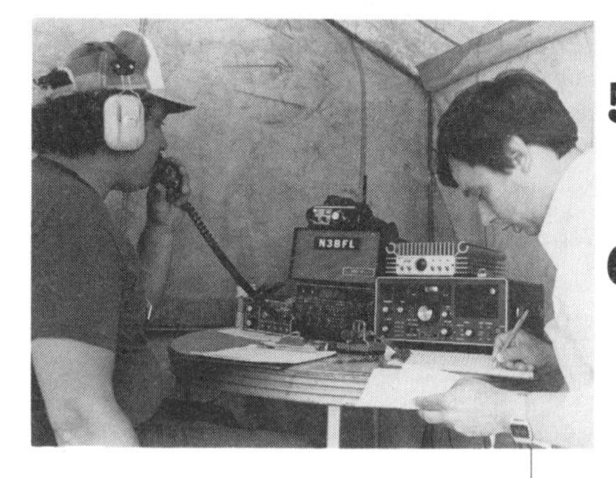

KDØLQ
ST. LOU

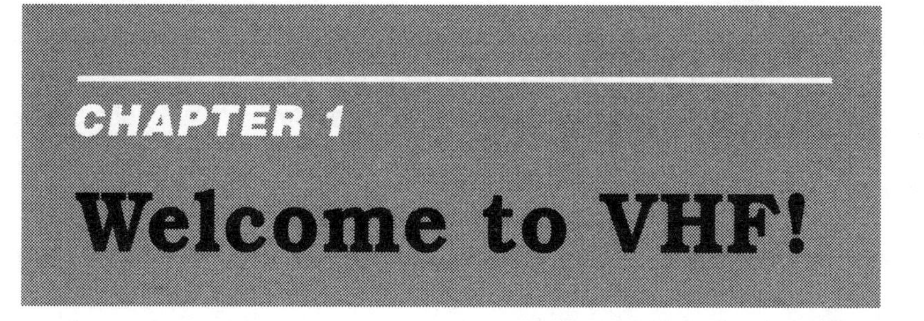

CHAPTER 1
Welcome to VHF!

By Steve Ford, WB8IMY

What brings you to VHF? Are you a newly licensed amateur who is interested in the VHF bands? Or are you a veteran in search of something new to rekindle your interest? In either case, your curiosity is about to be amply rewarded!

Some hams think strictly in terms of how far they can throw their signals. When they look at the VHF bands, they see nothing but limitations. The true VHF enthusiast knows different! In many

Dave Patterson, WB8ISZ, uses his 2-meter hand-held transceiver to direct the launch of a model rocket. *(photo by Steve Ford, WB8IMY)*

ways, it is HF that is limited. Pity the poor HF operator, crowded into less than 4 MHz of spectrum from 160 through 10 meters. We VHFers enjoy 41 MHz of wide-open territory in several bands from 50 to 450 MHz!

Imagine the plight of the hapless HF operator who needs reliable *local* communications. He has to search for a relatively quiet spot on 40 or 80 meters and hope that his friend across town will hear him! On VHF it's as easy as picking up a hand-held transceiver.

When it comes to mobile operating, VHF is ideal. VHF transceivers and antennas are compact. HF rigs, on the other hand, can require substantial room and the antennas tend to be large and...well, somewhat less than sleek! If a VHF operator needs to access the HF bands, a remote base or a repeater with an HF link can do in a pinch.

VHF operators can connect their computers to vast networks through *packet radio*. The HF operator can use packet too—at an excruciatingly slow 300 bits/second. If there's a burst of noise or a momentary fade, his station must retransmit the same packet again and again. (You have to admire that kind of patience and persistence!) On VHF, our packet networks function at a *minimum* data rate of 1200 bits/second—and often accelerate to 9600 bits/second or more. We don't have as much interference to overcome as do HF packet operators. Except in the most crowded conditions, our packet messages travel quickly and efficiently.

The HF operator has limited access to space communications through a couple of satellites. VHF, on the contrary, is a hotbed of satellite activity! You have an entire menu of satellites at your disposal. You can talk to astronauts and cosmonauts. You can relay packet messages through low-earth-orbiting satellites. You can receive *images* from satellites and you can even work DX from the high-flying Phase 3 birds! Try and top *that* below 30 MHz!

Only VHF operators can exchange live video. It's called

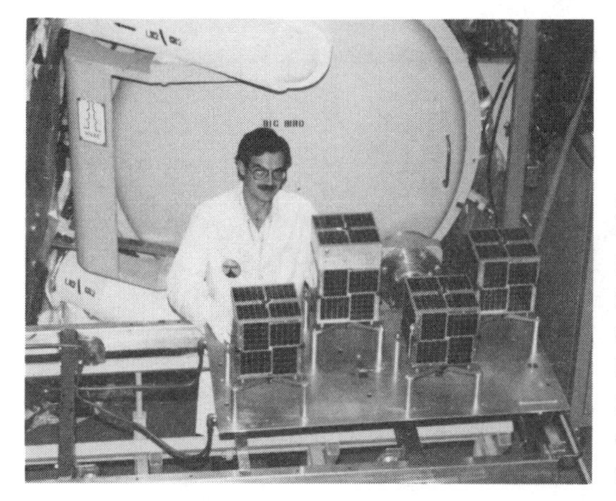

Jose Machado, LU7JCN, prepares to test the Microsats in a thermal vacuum chamber prior to launch. These satellites use the 2-meter and 420-MHz bands for uplinks and downlinks.

fast-scan TV, otherwise known as *ATV*. On ATV you'll see live images from other amateurs, just like conventional television. Some ATV enthusiasts are even placing TV cameras aboard high-altitude balloons that travel to the edge of outer space. Just imagine watching live video from a camera dangling 100,000 feet overhead!

On VHF you can explore the exciting world of weak-signal SSB and CW DXing. When it comes to working the weak ones, the rule of "the higher, the bigger, the better" applies to both HF and VHF antennas. On VHF, however, the antennas are considerably lighter and smaller. Hunting VHF DX is a fascinating experience. You tune slowly through the band, listening for distant signals in the gentle hiss of natural background noise. Compare that calm scenario to HF DXing, where you strain to hear the station through a din of competing signals.

In this book we'll explore the fascinating world of VHF and highlight several of the most active communication modes available. We'll even give you a taste of what exists *beyond* VHF in the exotic world of UHF and microwaves. Before we do anything else, however, let's take a quick look at the VHF bands.

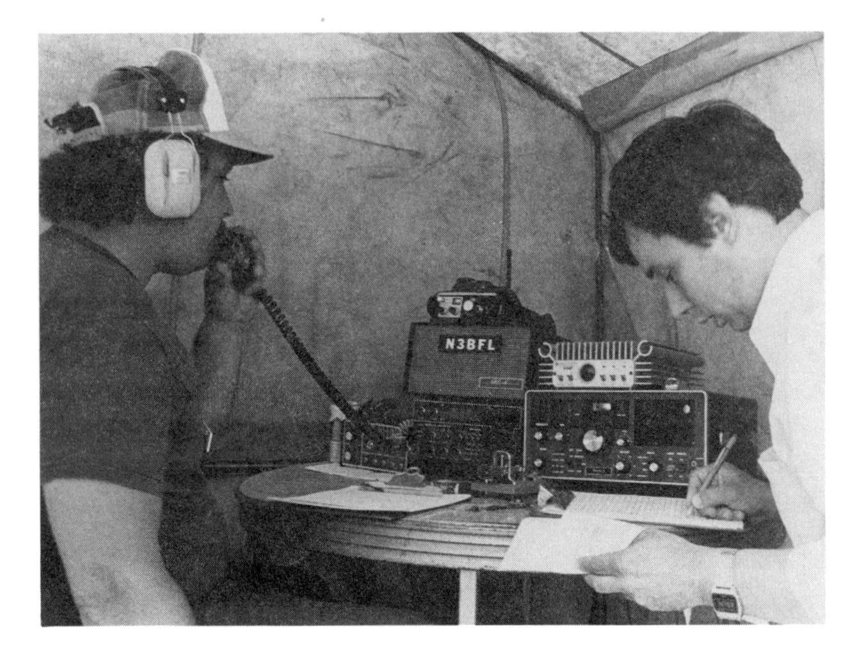

N3BFL, with W3PM logging, operates 2-meter SSB during Field Day.

6 Meters: The *HF* VHF band

The 6-meter band extends from 50 to 54 MHz and is open to all amateur license classes except Novice. Six meters straddles the line between what we consider to be HF and VHF. With its unusual position in the electromagnetic spectrum, 6 meters tends to have a split personality!

When the band is in its "VHF mode," it behaves much like 2 meters, offering excellent local and regional coverage. With modest CW or SSB equipment, small antennas and a decent location, you can expect to routinely work stations a few hundred miles distant. FM is also popular on 6 meters. Point-to-point *simplex* QSOs are common and repeaters are often available to provide coverage over wide areas.

Occasionally you'll have an opportunity to take advantage

Gus, HC2FG, is a 6-meter DXer and regularly works the United States from his station in Guayaquil, Ecuador.

of *meteor scatter* propagation. As the name implies, this involves bouncing your signal off the ionized trails of meteors as they plunge through the atmosphere (see Fig 1-1). This type of propagation is common on 6 meters and not as difficult as it sounds. You don't need to know the trajectory of the meteor and you don't need exotic equipment or antennas. Just listen for strong signals that seem to suddenly burst out of nowhere. You have to pounce quickly! Like the wispy trails themselves, the strong signals soon vanish.

Six-meter signals are also reflected off the polar auroras.

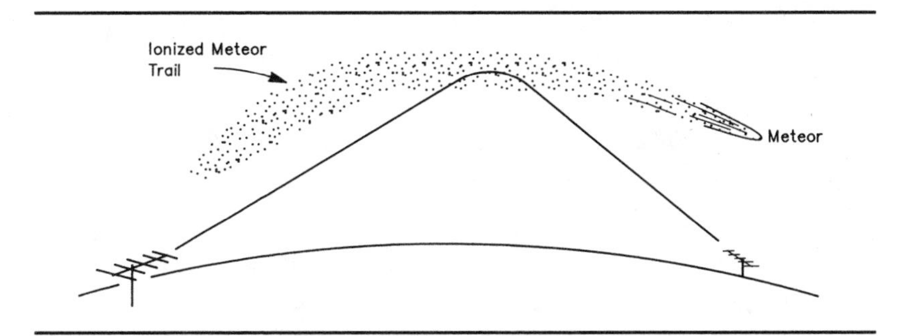

Fig 1-1—VHF radio signals can be reflected off the blazing trails of meteors as they enter the atmosphere, opening brief DX paths for vigilant operators!

This unusual mode of propagation can provide excellent VHF DX. Auroral signals have a distinct, distorted sound that you'll always recognize after you've heard it the first time. SSB can be used despite the distortion. CW is also very popular.

Six meters can enter its "HF mode" at virtually any time, but it's highly influenced by the seasons and sunspot cycles. Activity is highest in the summer and fall, but the band can explode with continental or worldwide DX when you least expect it!

Tropospheric band openings (or simply *tropo*) are the products of weather—usually slow-moving high pressure systems. A large tropo opening can spread your signal over half the continent, but such opportunities are rare on 6 meters.

When *sporadic E* (often called "E_S" for "E sporadic") comes into play, 6 meters really gets cooking! You can expect sporadic E on most days during June and July. December and January are active months, too. A typical E_S band opening will allow you to reach most areas of North America with ease. If you live near the East or West coasts, strong E_S openings have the potential to bring signals from other continents!

Sporadic E propagation can last from several minutes to several hours. During this time the propagation can shift from one geographic area to another. Stay on your toes and monitor the band frequently. Automated 6-meter beacons operate 24 hours a day. Listen for them and you'll get a pretty good idea of 6-meter band conditions at any given time. (You'll find a listing of beacons for *all* VHF and UHF bands in the *ARRL Repeater Directory* and *The ARRL Operating Manual*.)

When ionospheric propagation is particularly strong, the Maximum Usable Frequency (or *MUF*) will reach the 6-meter band. Under these conditions, 6 meters behaves very much like 10 meters. You won't need high power or sophisticated antennas to work choice DX. If the propagation path is carrying your signal into Europe, 10 watts is almost as good as 1000 watts. Most

6-meter DX is worked on SSB, although CW and even FM repeaters get into the act, too!

Just how far *can* you go on 6 meters? Well, the current distance record is over 12,000 miles. Not bad for a VHF band!

2 Meters: From Moonbounce to Repeaters

Of all the VHF bands, the 144- to 148-MHz segment known as 2 meters is the most popular. Like the 6-meter band, 2 meters is available to all amateurs except Novice licensees.

When FM repeaters began to appear in the late '60s and early '70s, 2 meters was the band of choice. It offered plenty of room for activity and its propagation characteristics were ideal for local repeater systems.

Over the past two decades, 2-meter FM repeaters have spread throughout the country. In highly populated areas, almost all available repeater frequencies are occupied. Many repeater systems, both rural and metropolitan, offer an *autopatch* function which permits hams to make telephone calls via the repeater. More sophisticated systems provide a wide array of features such as links to other bands. (Imagine using your 2-meter hand held transceiver to talk to Europe on 10 meters!) Repeaters

A compact VHF transceiver fits perfectly in the dashboard of this automobile. (*photo by Kirk Kleinschmidt, NTØZ*)

are also linked with each other to form large *integrated* networks. One network system known as the Evergreen Intertie covers a large area of California, several northwestern states and parts of Canada.

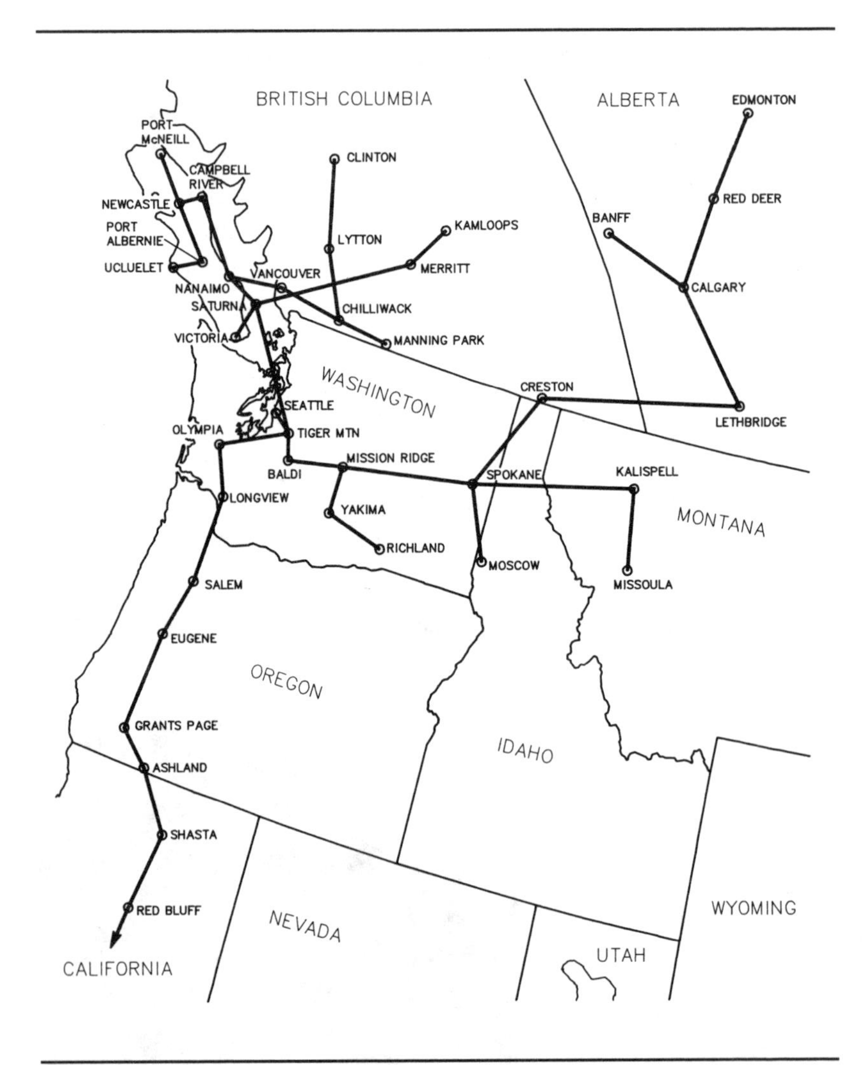

Fig 1-2—A partial map of the Evergreen Intertie, a system of integrated voice repeaters that allows long-distance communications with low-power hand-held transceivers.

Although repeater activity seems to dominate the band, there is much more to enjoy on 2 meters. Packet radio is very popular from 144.91 through 145.09 MHz; its rapid growth has been similar to the FM repeater phenomenon. Traditional packet frequencies have become too crowded in some areas, forcing packet systems to move to the segment between 145.50 and 145.80 MHz.

There is also quite a bit of satellite activity on 2 meters, mostly between 145.50 and 146 MHz. In this area of the band you'll find *uplinks* and *downlinks* to various satellites. The Russian *MIR* space station is often heard on 145.55 MHz and US space shuttle astronauts can occasionally be found in this portion of the band as well.

At the lower end of the band—from 144 to 144.5 MHz—you'll find CW and SSB enthusiasts. Like their 6-meter counterparts, they're DX chasers, constantly hunting for elusive band openings.

Two-meter band openings are often provided by tropospheric propagation. Good tropo conditions can open the band over huge areas of the country, sometimes for days at a time. When it comes to tropospheric

A 2-meter ground-plane antenna is ideal for working local voice repeaters and packet stations. The antenna is inexpensive and easy to mount on a chimney or other convenient support. (*photo by Steve Ford, WB8IMY*)

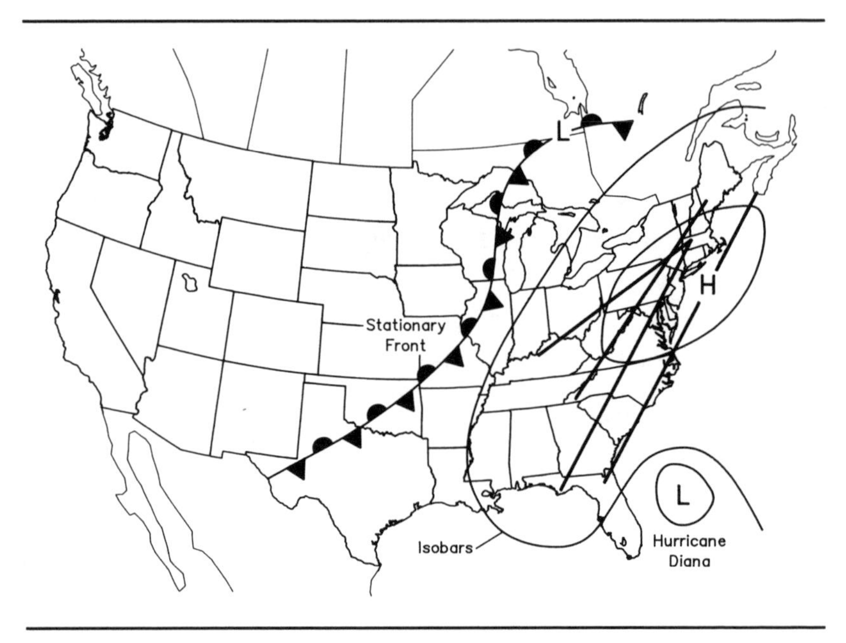

Fig 1-3—This is a classic example of the type of weather conditions that cause tropospheric band openings. Notice the large, slow-moving high-pressure system centered over eastern Massachusetts. The dark lines indicate 2-meter signal paths observed during the band opening.

propagation, 2 meters has a decided advantage over 6 meters.

Sporadic E propagation also occurs on 2 meters, but is much less common than on 6 meters. A good E_S opening on 2 meters will provide communications over a thousand miles or more. There are exceptions to every rule, however. A strong sporadic E event can extend 2-meter CW and SSB coverage out to several thousand miles!

Two-meter weak-signal operators also like to bounce their signals off distant objects—such as meteor trails, polar auroras and even the moon. These objects act like mirrors, reflecting signals in new directions and opening pathways for long-distance communications.

Two meters is the highest VHF band that offers frequent

meteor-scatter opportunities. QSOs over distances of hundreds or thousands of miles are not uncommon. During an active meteor shower, you may hear as many as 50 signal bursts—*pings* in meteor-scatter parlance—per hour! Two-meter signals can also be reflected off the polar auroras, with CW being the preferred mode (auroral distortion makes the use of SSB nearly impossible at 2 meters and above).

Finally, there is that great bullseye in the sky—the moon. Moonbounce (also called *EME*—for *earth-moon-earth*) is very popular among 2-meter weak-signal aficionados. It was once assumed that only stations with large, high-gain antennas and enormous output power could work moonbounce. Recent developments have proven that this is not always the case. Moonbounce contacts have been made with a single multi-element beam antenna and less than 100 watts! Moonbounce is one type of 2-meter activity that offers the opportunity to chase international DX. As long as you and the other station can see the moon simultaneously, a moonbounce contact is possible. Amateurs in the US often work stations in South America and Europe via the moon. Since the returning signals are so weak (after traveling to the moon and back, who wouldn't be?), CW is the mode of choice.

222 MHz: 2 meters' "Higher Cousin"

Although the frequency is nearly double, the 222-MHz band shares many characteristics with 2 meters. This band is open to all license classes—*including Novice*. However, Novices are restricted to 222.1 to 223.91 MHz and 25 watts PEP output.

222 MHz is home to a number of FM repeaters, especially in metropolitan areas. Novices make good use of their privileges on these machines! Many 222-MHz repeaters have crossband links to 2 meters and other bands as well. In terms of coverage, there is no fundamental difference between a 2-meter repeater and a

Here's a 222-MHz mobile antenna you can build yourself. Check the *ARRL Handbook* for details.

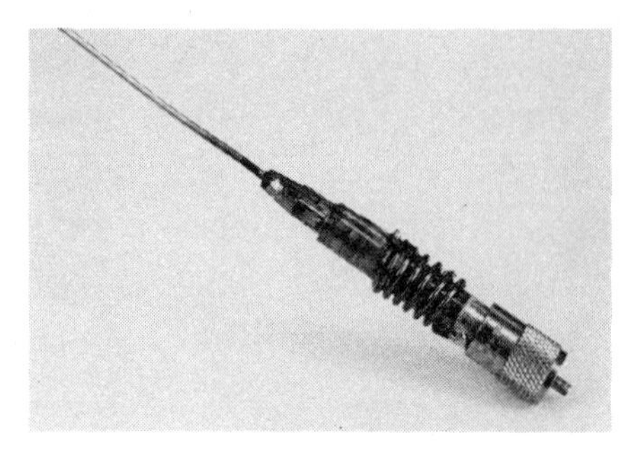

222-MHz repeater. If both systems are well engineered, they should provide equal coverage.

There is also packet activity on 222 MHz. Bulletin boards may be found in some areas, although much of the activity is in the form of *backbone* systems that relay traffic from one bulletin board or network system to another. As more Novices and Technicians discover 222 MHz, other types of packet operations are almost certain to increase.

Weak-signal SSB and CW operators find plenty to enjoy on 222 MHz—particularly during contests. Just as FM enthusiasts discover similarities between 222 MHz and 2 meters, the same is true for weak-signal operators. With a little patience, and an ear to the propagation beacons, 222 MHz can produce many surprises!

Tropospheric propagation is relatively common on 222 MHz, occasionally allowing communication over several hundred miles or more. *Radiation inversion* is a form of tropo propagation that creates some intriguing conditions during the summer months. In the twilight hours when the air is still, weak-

This impressive station is active on all VHF bands—especially 222 MHz! Donna, N2FFY, enjoys hunting VHF DX.

signal operators on 222 MHz often notice a gradual im-provement in signals out to a range of about 200 miles. Then, with the onset of night, the range suddenly decreases. That's the intriguing characteristic of radiation-inversion propagation!

Sporadic E is rare on 222 MHz, but not entirely unknown. On the other hand, meteor scatter and aurora contacts produce their share of DX excitement! Moonbounce is possible on the band, but activity is low.

420 MHz: The VHF/UHF Borderline

Is the 420- to 450-MHz band (commonly known as 70 cm) VHF or UHF? Technically speaking, it's a UHF band, but 70 cm shares many propagation characteristics—and activities—with the VHF bands.

Hams in the United States share the 420-MHz band with other services on a *secondary* basis. This means 420-MHz enthusiasts may be affected by signals from other services, or may have to modify their operating so as not to cause interference to these services. Amateurs living near designated military bases and other areas with high concentrations of scientific or defense-related activities are restricted to a maximum output of 50 watts PEP. In addition, hams may not operate in the 420- to 430-MHz portion of the band if they live within a narrow zone that stretches along the US/Canadian border. See the *FCC Rule Book* for specific details.

Despite the restrictions, it's important to remember that the 420-MHz band is *huge*! The band encompasses a total of *30 MHz*. That's enough spectrum to hold everything from the AM broadcast band all the way through 10 meters with room to spare! All amateurs except Novice licensees have full privileges on 420 MHz.

420 MHz is the lowest band where fast-scan television (ATV) is permitted. ATV has been at home on 420 MHz for many

Put eight 432-MHz beams together and what do you get? A moon-bounce array with a *very* powerful signal. The proud owner of this antenna system is W9IP/2.

years, making it the most active band for amateur television. A natural choice for ATV, the 420-MHz band has ample room for wideband video signals. Since a television signal spreads its transmitted energy over several megahertz, high-gain beam antennas are required for most ATV stations. To give more hams an opportunity to enjoy this mode, ATV repeaters have appeared in many areas.

ATV is also popping up in some unusual places! A number of amateur groups and clubs are launching high-altitude balloons equipped with ATV cameras. A typical balloon carries its payload to about 100,000 feet where the low atmospheric pressure causes the balloon to expand and burst. If everything goes as planned, a parachute lowers the ATV payload gently to the ground.

The video images transmitted from these balloon flights can be spectacular. If high-resolution cameras are used, ground details can be recognized easily. At 100,000 feet, the curvature of the earth's horizon is visible and stars can be seen in the cold blackness beyond. The popularity of high-altitude ATV balloons is increasing rapidly. Some ambitious plans include remote-controlled gliders, various types of balloon-launched ATV *rockets* and long-duration ATV balloon flights.

420 MHz is also the most popular band for amateur satellite activity. In fact, 435 through 438 MHz has been allocated for amateur satellites *exclusively*. That's as large as the entire 222-MHz band! You can work international DX through AMSAT-OSCAR 13 via *mode-B*—a 435-MHz uplink and 145-MHz downlink. Packet mail can be picked up and delivered through Fuji-OSCAR 20 or the Microsats. You can even watch video images of the earth as glimpsed from OSCAR-18 or OSCAR-22. When you consider all the satellite activity going on today, it's easy to understand why the 420-MHz band plays such an important role.

FM repeater activity is very common on 420 MHz, but this was not always the case. Two-meter repeater operators initially used the band to establish links between remote receivers and the main repeater site. The remote receivers listened on 2 meters and relayed to the repeater via dedicated 420-MHz links. With remote "ears" in the proper

Blast off! A model rocket becomes a blur of speed as it leaves the launch pad. Its nose cone contains a miniature 420-MHz telemetry transmitter. (*photo by Steve Ford, WB8IMY*)

Amateur video from the edge of space! Helium-filled balloons carry ATV cameras and transmitters to 100,000 feet, providing amazing views like this to hams below. (*photo by Bill Brown, WB8ELK*)

locations, a 2-meter repeater system could offer excellent coverage to its users. This arrangement has been proven to be very effective and is still in use today.

Even by the early '70s, the 2-meter repeater subband was becoming crowded, prompting a gradual influx of full-fledged repeater systems in the 420-MHz band. (The 222-MHz band may have seemed like a more logical choice for expansion, but 222-MHz FM equipment was scarce at the time.) Although many FM operators still favor 420 MHz, they often maintain remotely controlled links that allow their 420-MHz repeaters to receive and transmit on 2 meters as well. Talk about having your cake and eating it too!

While we're on the subject of links, it's worthwhile to mention that the 420-MHz band supports a fair amount of packet activity in the form of backbone links—like those on the

222-MHz band. As 2-meter packet frequencies become more crowded, you can expect to see further packet expansion at 420 MHz.

SSB and CW enthusiasts have plenty of room to operate in the 420-MHz band. Although the band presents some challenges to working DX, the antennas are considerably smaller than those required on the other VHF bands. A high-gain antenna need only be a few feet in length. With a little work, these small antennas can be combined into powerful, multi-antenna arrays!

Moonbounce, meteor scatter and aurora all provide DX opportunities on 420 MHz. Tropospheric propagation is also common on the band. When weather conditions favor a strong tropo opening, your signal can span several hundred miles. In fact, the 420-MHz band enjoys more tropospheric enhancement than the lower-frequency bands.

Of all the bands discussed thus far, 420 MHz is most prone to signal attenuation caused by topographic features (such as hills and mountains). If you intend to chase DX on 420 MHz, you'll need to get your antenna as high off the ground as possible. Feed-line loss is also higher at 420 MHz so you'll need to use low-loss cable such as Belden 9913 or lower-loss *hardline*.

If you don't have a tower, don't ignore 420 MHz! You can still operate via satellites and repeaters—including ATV repeaters. In addition, nearby hills and mountains offer excellent locations for VHF contesting. Take your portable equipment to the summit and make a camping trip part of your contest. With a hill or mountain as your "tower," you'll be surprised at how many contacts you can make!

Are you sold you on the excitement of the VHF bands? Good! We've spent a number of pages speaking in generalities about what you can do on these fascinating frequencies. Now it's time to talk in detail about the most popular modes and activities...beginning with FM and repeaters.

CHAPTER 2

FM and Repeaters

By Steve Ford, WB8IMY

Everyone is on FM—at least it seems that way on some VHF bands! You can't go to a major hamfest without seeing hundreds of hand-held transceivers dangling from belts and pockets. (The attendees often look like Dodge City gunfighters gathering for a big shoot-out.) And when you're walking through the hamfest parking lot, you see mobile antennas of every description, most attached to FM rigs. What caused this incredible fascination with VHF-FM?

Repeater Magic

More than anything else, *repeaters* are the driving force behind the popularity of VHF-FM. If you love technology—and you wouldn't be a ham if you didn't—FM repeaters seem almost magical. Thanks to repeaters, hams with low-power transceivers can chat with other hams many miles away. Some of these conversations would be impossible without their assistance—no matter how much power was being used. How do repeaters perform such amazing tricks?

A repeater is a *duplex* device. That is, it transmits on one frequency while *simultaneously* listening on another. Transmitting and receiving at the same time is a neat trick by itself. Some repeaters do this by using separate antennas for the transmitter and receiver. The majority accomplish the same task using only *one* antenna! In this instance, the receiver and transmitter are connected to a specially tuned filter known as a *duplexer*. The duplexer keeps the transmitted and the received signals separate as they flow to and from the same antenna. Without a duplexer to keep the peace, a single-antenna repeater would have a very short lifespan!

Repeaters are usually installed at the highest possible location—on a tall building, a water tank, a broadcast tower, or at the summit of a hill or mountain. From its lofty vantage point, the receiver can pick up signals from a very wide area. At the same time, the transmitter can be heard to a range of 10, 20, 30 miles or more.

With repeaters to lend a helping hand, base and mobile operators can enjoy reliable VHF-FM communications. All they have to do is transmit on the repeater *input* frequency and listen on the *output* frequency. When the repeater hears a signal on the input, it *instantly* retransmits (repeats) the signal on the output (see Fig 2-1). An input/output frequency separation of 600 kHz is the 2-meter standard. On 222 MHz, a 1.6-MHz separation has been established and on 420 MHz, 5 MHz is the standard.

Almost every populated area of the United States is within range of at least one FM repeater. (Quite a few nonpopulated areas are covered, too!) Many of these repeaters offer sophisticated features such as autopatches and crossband links. Two meters and 420 MHz are the most active FM bands, although you'll also find FM enthusiasts on 6 meters and 222 MHz. The 222-MHz band deserves extra attention since it is the only VHF band that allows *all* licensees to use FM repeaters. As a result, activity has been growing on 222 MHz in recent years.

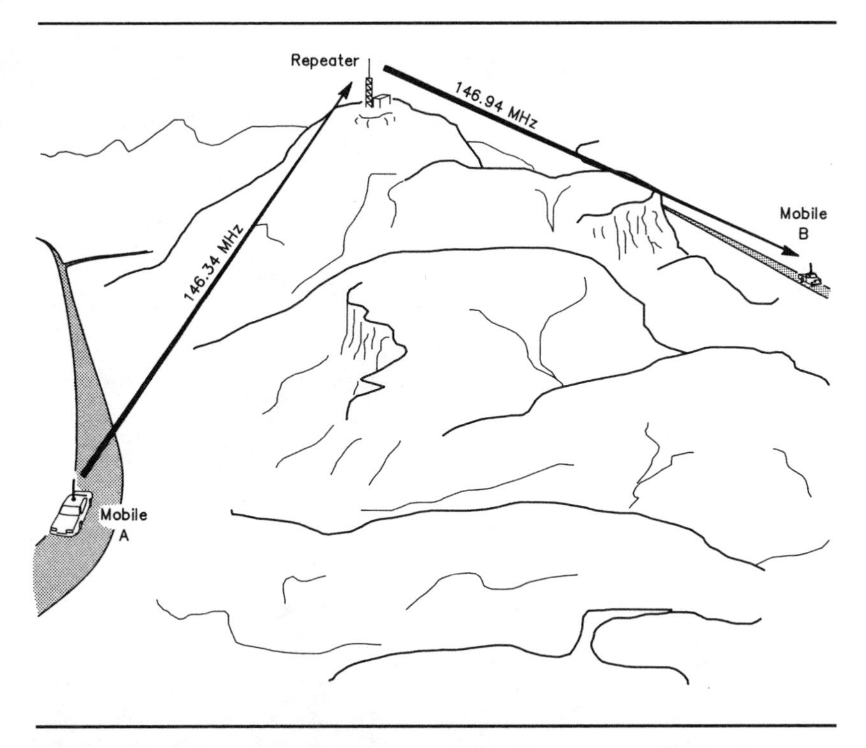

Fig 2-1—The mountain blocks direct (simplex) communication between mobile stations A and B. When mobile A transmits on 146.34 MHz, however, the repeater detects its signal and relays it to mobile B on 146.94 MHz. When it's mobile B's turn to transmit, the same signal relay occurs in reverse.

Although it's not VHF, 10 meters offers FM opportunities in the upper portion of the band. You'll find repeater outputs from 29.620 to 29.680 MHz. The national simplex calling frequency is 29.600 MHz. When propagation conditions are favorable, you can work plenty of DX on 10-meter FM! A number of VHF repeaters provide crossband links to this fascinating band—as we'll discuss later.

It's obvious that repeater operating has practical advantages, but that doesn't fully explain the continued popularity of FM.

With hand-held in hand, Max Trescot, K3QM, prepares to board the Goodyear blimp for a little sightseeing and "blimp mobile" operating. With several thousand feet of altitude, a few watts goes a long way!

When you get down to the bottom line, the answer is clear: VHF-FM is *fun*!

Enjoying FM

How many ways can you use VHF-FM? The list is endless! Are you going on a hiking expedition with your Amateur Radio friends? Take your hand-held transceivers along! Everyone can walk at their own pace and still keep in touch. The same idea applies to bicycle excursions, fishing trips and other outdoor activities.

Perhaps you're helping a ham friend move into a new home. Wouldn't it be great to use VHF-FM to coordinate your activities?

"WA7GHR from NT7Z"

"NT7Z from WA7GHR. Where are you?"

"I'm on the interstate and I'll be there in 15 minutes. I have your sofa and refrigerator. Better get the driveway clear."

"Will do. By the way, you'll be happy to know I have lunch waiting for you."

"Perfect! I'll call you again when I turn onto your street. NT7Z clear."

When Dave, WB8ISZ, and I launch model rockets, we always take our hand-helds. If a rocket lands too far from the launch site, someone has to track it down. When I'm the tracker, I listen to Dave as he directs me to the landing area. I can also report the condition of the rocket when I finally find it.

As you can see, VHF-FM will add a new dimension to your other hobbies and interests. It's also ideal for public-service functions such as parades and "walk-a-thons." Many Amateur Radio clubs actively participate in local public-service events. Why not volunteer your time for a good cause? Not only will you enjoy yourself, you'll sharpen your operating skills as well.

Using an FM Repeater

The first step in using a repeater is finding one! The *ARRL Repeater Directory* is a handy reference that will help you locate repeaters in your area. Another approach is to use the scanning functions built into many modern FM rigs. By simply scanning the band and listening, you should discover several repeaters.

A repeater is usually referred to by its output frequency. If we were operating on 2 meters and I said, "I'll meet you on the 94 machine," I mean that I'll meet you on the 146.94 repeater. On most FM transceivers, the frequency shown on the display is the *receive* frequency. To meet me on the repeater, all you'd have to do is tune your rig until 146.94 MHz appeared on the display. Since we're using a 146-MHz repeater in our example, you'd also need to make sure the transceiver was set to transmit 600 kHz *below* the receive frequency, not above! (On the other hand, 2-meter repeaters operating above 147.00 MHz use input frequencies 600 kHz *above* their transmit frequencies.) Finally, you'd want to verify that your transceiver was in the *duplex* mode (transmitting on one frequency, receiving on another).

My 2-meter FM transceiver is about ten years old, so whenever I change repeaters I need to check all my switches and make certain that everything is ready to go. Many newer models include microprocessor-controlled memories that allow you to preprogram the rig for your favorite repeaters—including the proper frequency offsets used by each one. Changing from one repeater to another is often a matter of pressing a single button.

Your First Transmission

Let's assume that you've found a strong repeater and you've set your transceiver to the proper frequencies. Now you want to talk to someone! If the repeater is quiet, hold the microphone close to your lips and press the push-to-talk switch. Simply say, "NØLEI listening" or "NØLEI monitoring." If someone on the repeater wants to chat with you, they'll respond. If you hear nothing, wait a couple of minutes and try again.

Whatever you do, don't call "CQ" on a repeater! Calling CQ is fine for simplex contacts, but it's considered bad operating technique on repeaters. In addition, avoid saying, "NØLEI listening 146.94." There's no need to announce the frequency. You know which repeater you're using and so does everyone else!

If someone responds to your call, the rest is easy. Just relax and enjoy yourself. Tell the other station who you are and where you're located. If the repeater isn't hearing the other station well (if there is noise on his signal whenever he transmits), you'll want to let him know.

During your first few transmissions it's normal to feel a little nervous and tongue-tied. If you find yourself in this position, just throw out a question to the other person. Ask what they do for a living. Ask what other bands or modes they operate. People love to talk about themselves and you may find that you have interests in common. Listen to the answers and comment on what the other person says. This is the fine art of conversation—whether it's in person or on a repeater.

Timers and Courtesy Beeps

As you become more comfortable on the air, you may find that your transmissions are getting longer and longer. Beware! Repeaters have built-in timers that limit how long one station is permitted to transmit without a break. The timer settings can vary from one repeater to another. Some are as long as five minutes while others are as short as 60 seconds! (Three to five minutes is the norm.) If you talk so long that the timer runs out, the repeater will shut down completely until you stop transmitting. This is known as *timing-out* a repeater.

Timing-out a repeater can be a little embarrassing. Some repeaters use a warning system to let the users know when someone is about to exceed the limit. (Too bad the transmitting station can't hear it!) On one of my local machines, a synthesized voice announces, "Repeater time-out!" just before it shuts down. When the station finally stops transmitting and the repeater returns to the air, the voice says, "Repeater time-out canceled!" The long-winded operator hears *that* chastising announcement loud and clear. Hams are good-natured people, though. Most have timed-out a repeater at least once in their lives and they know how it feels!

Some repeaters also use *courtesy beeps*. The purpose of a courtesy beep is to force a delay between the end of one transmission and the beginning of another. This delay allows another station to join the conversation, or interrupt with emergency traffic. If you're using this type of repeater, you'll hear the beep a second or two after the end of your transmission. If no one breaks in before the beep, the other station is free to begin transmitting.

Repeaters usually impose a penalty for ignoring the courtesy beep. Every time the beep sounds, it resets the time-out timer. If you're too quick on the trigger and transmit before the beep, the timer keeps running. Let's say the timer was set for three minutes and the other station spoke for two minutes and forty-five seconds. If you begin your reply before hearing the beep, you'll only enjoy about 15 seconds on the repeater before it shuts down!

Joining a Conversation

Another way to make new friends on a repeater is to join a conversation in progress. Listen to the discussion for a while and see if there is anything you can contribute. If so, wait for one station to stop transmitting. Before the next station begins, quickly press your microphone button and say your call sign once.

Jim Jones, N9DIX, uses his FM mobile rig to provide vital communications in the aftermath of a devastating tornado in Lemont, Illinois. (*photo courtesy of Marty Barton, KA9ZJJ*)

Avoid using the word "break" or "break-break." Among some repeater groups this means you have emergency or priority traffic.

If the other station hears your call sign, he or she will either acknowledge you and continue, or stop abruptly and turn the conversation over to you.

"I can't wait to try this new rig at home. With 45 watts output, I should be able to reach you on simplex, Tom."

"AA8YR"

"AA8YR acknowledged. Well, you'd better check your old power supply before you hook it up. I'm not sure it'll supply enough current to operate the transceiver at the 45-watt level. Let's see what the new station has to say. AA8YR from WB8QVC."

As you spend more time on a particular repeater, you'll begin to develop on-the-air friendships. Hams who know each other tend to congregate on repeaters in the mornings and evenings when they're driving to and from work or school. Sometimes a loose conversation that begins between two stations can quickly turn into a multi-station *roundtable*, with each ham passing control to whomever seems to be next in line. Roundtable QSOs are fun and they can make a long commute seem quite a bit shorter!

Closed Repeaters

The vast majority of repeaters you'll encounter will be *open*. That is, they're open for use by anyone. There are some repeaters, however, that restrict access for various reasons. These are known as *closed* repeaters.

The most popular method for regulating repeater access is through the use of a *continuous tone-controlled squelch system* or *CTCSS*. To use this system, your transceiver must be equipped with an encoder that sends a low-frequency tone every time you transmit. The tone is too low for other stations to detect, but a properly equipped repeater receives it very well. As you might guess, the repeater will not activate unless it hears the correct tone.

Some repeaters have been forced to use CTCSS because of interference on their input frequencies. One repeater may occupy the same input and output frequencies as another, for example. Even though the repeaters are separated by a considerable distance, the users of one repeater are often heard by *both* systems. This creates an annoying situation for everyone concerned! If neither repeater can move to another frequency, the use of a CTCSS may be the only solution.

This type of CTCSS-controlled repeater is not closed in the true sense of the word. The group or club that owns the repeater may openly publish the tone frequency for use by members and non-members alike. Many repeater groups list their CTCSS tones in the *ARRL Repeater Directory*.

There are other repeaters that are truly closed systems. Their CTCSS tones are carefully guarded secrets and only members in good standing have access to the information. Some closed repeaters are maintained through the financial support of small groups. Since the members have invested a substantial amount of time and money in their repeaters, they may want to keep them "private." Other repeaters are closed because they serve the needs of groups with common interests. Although closed repeaters violate the spirit of Amateur Radio, they are not illegal.

The Autopatch: Your Telephone Connection

An autopatch is a device that allows hams to make telephone calls through repeaters. In most repeater autopatch systems, the user simply has to generate the standard telephone company tones to access and dial through the system. This is usually done by interfacing a telephone tone pad with the user's transceiver. Dual-tone multifrequency (DTMF) tone keypads are often mounted on the front of hand-held portable transceivers, or on fixed or mobile transceiver microphones.

There are a number of important rules that govern autopatch operation. See the ARRL Autopatch Guidelines sidebar for a thorough treatment of the subject.

If you have a legitimate reason to use the autopatch, how do you do it? First, you must access (turn on) the autopatch, usually by pressing a designated key or combination of keys on the tone pad. When you join the group or club that operates the repeater, they'll provide you with a list of autopatch access codes.

Make sure the frequency is clear before you begin and remember to identify your station. It's common to say, "WB8ISZ accessing the patch," before you transmit the code sequence. When you hear a dial tone, you'll know you've successfully accessed the autopatch. Now, punch in the telephone number you wish to call.

When you establish the call, let the person know that it's not a private conversation. If you just say, "Hi, I'm talking to you from my car," they might think you have a mobile cellular telephone. They should also be told that they have to wait their turn to talk. Many autopatches have timers that terminate the connection after a certain period of time, so keep your telephone conversation short and end the autopatch as quickly as possible.

The procedure for turning off the autopatch is similar to the procedure for accessing it. A code must be entered to return the repeater to normal operation. Don't forget to identify when you finish. ("WB8ISZ clearing the patch.")

Autopatches are very handy in a variety of situations. If I'm supposed to meet someone and I'm running late, I can call ahead and let him know. If I'm having car trouble, I can call one of my friends and ask them to pick me up.

Autopatches are put to their best use during emergencies. If you see an accident along the highway, for example, your ability

ARRL Autopatch Guidelines

Radio amateurs in the US enjoy a great privilege: The ability to connect their repeaters with the public telephone system.

As with any privilege, this one can be abused, and the penalty for abuse could be the loss of the privilege for all amateurs. What constitutes abuse of autopatch privileges? Without specific regulations governing its use, the answer to this question depends on one's perspective.

The important question facing amateurs is this: Should autopatches be subject to reasonable voluntary restraints, thereby preserving most of our traditional flexibility, or should we risk forcing our government to define specifically what we can and cannot do? When specific regulations are established, innovation and flexibility are likely to suffer.

It's the policy of the ARRL to safeguard the prerogative of amateurs to interconnect their repeaters to the public telephone system. An important element of this defense is encouraging amateurs to maintain a high standard of legal and ethical conduct. The guidelines here address these standards.

1) Autopatches involving the business affairs of any party must not be conducted at any time. (The FCC considers nonprofit and noncommercial organizations businesses and forbids the use of Amateur Radio to facilitate their day-to-day business in any way.) The content of an autopatch call should be such that it's clear to any listener that business communications are not involved. (Particular caution must be observed when calling any business telephone.) Calls to place an order for a commercial product must not be made, nor may calls be made to one's office to receive or to leave business messages. Calls made in the interests of highway safety, however, such as for the removal of injured persons from the scene of an accident or for the removal of a disabled vehicle

to use an autopatch could make a life-or-death difference. Whenever you use an autopatch to contact police or fire departments, make sure to tell the dispatcher *exactly* where you are located. Some 911 emergency telephone systems use a from a hazardous location, are permitted.

2) Autopatches should never be made solely to avoid telephone toll charges. Autopatches should never be made when normal telephone service could be used just as easily.

3) Third parties (nonhams) should not be put on the air until the responsible control operator has explained the nature of Amateur Radio. Control of the station must never be relinquished to an unlicensed person.

4) Autopatches must be terminated immediately in the event of any illegality or impropriety.

5) Autopatch facilities must not be used for the broadcasting of information of interest to the general public. If a repeater can transmit information of interest to the general public, such as weather reports, those transmissions must occur only when requested by a licensed amateur and must not conform to a specific time schedule. The retransmission of radio signals from other services is not permitted in the amateur service. (For example, you cannot retransmit a NOAA weather broadcast.) However, the retransmission of taped material from other sources is permitted. You could play back a tape recording of a friend reading instructions for wiring a microphone, provided you follow accepted amateur practices and identify your station appropriately.

6) Station identification must be strictly observed.

7) Autopatches should be kept as brief as possible, as a courtesy to other amateurs; the amateur bands are intended primarily for communication among radio amateurs.

8) If you have any doubt as to the legality or advisability of an autopatch call, don't make it.

Strict compliance with these guidelines will help ensure that our autopatch privileges will continue to be available in the future, which helps us contribute to the public interest. See the *FCC Rule Book* for additional information.

computer to trace the origin of every call. Since you're using an autopatch, the computer will indicate that your call is coming from the repeater site, miles from the emergency scene!

Tell the dispatcher where the emergency is located. ("It's on

I-91 northbound, about a mile south of exit 15.") If you've stopped to offer assistance, give precise details concerning the situation. If it's an auto accident, for example, how many vehicles are involved? Are there any injuries? How many lanes of traffic are blocked? You don't have to communicate every scrap of information, just the most important facts.

Want to Try Another Band? Use the Links!

Many sophisticated repeater systems allow their users to transmit and receive on more than one band. By entering the appropriate codes from your keypad, for example, you may be able to activate a *crossband link* from 2 meters to 10 meters. Whatever you transmit to the 2-meter repeater will be retransmitted on 10 meters. If someone responds on 10 meters, their signal will be relayed to you via the repeater on 2 meters. Crossband links to HF frequencies can bring exciting DX opportunities. I once enjoyed a conversation with an amateur in South Africa on 10-meter FM while sitting on my back porch using a 2-meter hand-held transceiver!

A repeater may also have more than one crossband link. Some repeater systems allow operators to use these links as they please. Other systems maintain strict controls, allowing only the repeater trustee or other responsible individuals to turn them on or off. In many cases, a link may be active on a continuous basis. A classic example is a 2-meter repeater with a continuous link to the 70-cm band. This permits users on 70 cm to chat with their friends on 2 meters without having to change rigs.

If you're listening to a crossband conversation, it isn't always obvious that one of the hams is not on the same band you're on! This can be a little confusing until you get used to it. If your repeater has a crossband link to 10-meter FM, for example, you may hear someone say, "N5TXX monitoring." Is N5TXX local or not? He could be using the repeater on 2 meters, or he might be coming in on the 10-meter FM link. Some repeaters provide a special tone that's transmitted periodically whenever a link is in use..On other systems, you have to make your best guess!

Integrated Repeater Systems

Another version of crossband linking takes place in integrated repeater systems. An integrated system is really a network of repeaters covering a large area. (See the map of the Evergreen Intertie in Chapter 1.)

Integrated repeaters usually use 222-MHz or 70-cm links to connect one repeater system to another. Let's say that you're in Cincinnati, Ohio and you want to talk to someone in Louisville, Kentucky. The distance is too great for one 2-meter repeater to cover. If the repeaters between Louisville and Cincinnati are part of an integrated system, however, you can use your keypad to activate the links and establish a path.

Integrated repeater systems are becoming very popular in many areas of the country. Not only are they convenient and fun to use, they have great potential for emergency communications.

Go Simplex!

Simplex is a fancy-sounding word for a direct contact on a single frequency. After you've made a contact on a repeater, it's always best to move the conversation to a simplex frequency, if possible. This keeps the repeater free for stations that really need its assistance to communicate.

Simplex communications offer a degree of privacy impossible to achieve on a repeater—and you don't have to worry about timers or courtesy beeps! When selecting a frequency, make sure it's designated for FM simplex operation. The VHF band is subdivided for specific modes of operation, such as satellite communications and weak-signal CW and SSB. If you select a simplex frequency indiscriminately, you may interfere

Studying by Simplex!

I was tuning through the 2-meter FM simplex frequencies one evening when I heard an unusual conversation.

"Okay, what about the Battle of Hastings?"

"Ah...1066, right?"

"Yeah, but who won?"

"I can't remember. Charlemagne?"

"No!" someone else shouted. "It was William. William the Conqueror!"

"Very good, Greg. WB8ITK from WB8SVN. Frank is really dropping the ball tonight. Should I give him another chance?"

"Sure."

"Okay. WB8OFR from WB8SVN. In what year did Henry the VIII become king?"

It didn't take long to figure out what was going on and I'm sure you've guessed it, too. A group of students were using FM simplex to study for their history exams. One station would ask the questions and everyone else would have a chance to respond.

I began hearing the study group on a regular basis. Of course, the subject area would change according to the exam. Sometimes they'd get tired of studying and would start swapping opinions on the latest music or whatever came to mind. By using a simplex frequency, their conversations were relatively private and they weren't monopolizing a repeater.

Just when you think you've heard every possible use of VHF-FM, something new appears. This sounds like a great idea—and certainly more enjoyable than sitting alone with a textbook!

with stations operating in other modes. Check the *ARRL Repeater Directory* to find the FM simplex frequencies for your favorite bands.

VHF-FM is also an acceptable mode for contesting in sweepstakes, sprints, Field Day and other events. The only stipulation is that contacts must be made without repeaters and generally conducted off the national simplex or calling frequencies. Watch for announcements in *QST* and bring that 2- or 6-meter FM rig out to a mountaintop and call "CQ contest" sometime!

Traffic Handling

Any message relayed via Amateur Radio is known as *traffic*. With their wide coverage capabilities, repeaters are excellent tools for local and regional traffic nets. (Nets are groups of amateurs who meet on specific frequencies at specific times.)

The procedures for handling traffic on a repeater are about the same as handling traffic anywhere else. However, there are two things unique to repeater traffic handling: time-out timers and only occasional use of phonetics.

When you relay a message by repeater, release your push to talk (PTT) switch during natural breaks in the message to reset the timer. If you read the message in one long breath without resetting the timer, the repeater may shut down in the middle of your message. If it does, you'll hold up the net—and earn a red face—while you repeat the message!

The use of phonetics and repetition is not necessary when relaying a message over a repeater. The only time phonetics are necessary is to spell out an unfamiliar word or words with similar-sounding letters. (An exception: The word "emergency" is always spelled out when it appears in the preamble of a formal radiogram.) On the rare occasion that a receiving station misses something, she can ask you to fill in the missing information.

The efficiency provided by repeaters has made repeater traffic nets popular. There are probably one or more active repeater traffic nets in your area (consult the latest edition of the *ARRL Net Directory*). If there isn't a repeater net in your vicinity, why not fill the need by starting one yourself? Contact your ARRL Section Manager (SM) or Section Traffic Manager (STM) for details. *QST* lists your local officials; check page 8 or Section News in a current issue.

FM Transceivers—The Choice is Yours

Technology and new privileges have ushered in a Golden Age of FM rigs with advantages unheard of only a few years ago. You get what you pay for (some things never change), but you needn't smash your piggy bank just to get on the air. Cruise the flea markets with a few dollars and find an older FM transceiver.

Not long ago, all VHF rigs were monobanders. Today, dual-banders are becoming more popular. These transceivers offer 2 meters and 222 MHz, or 2 meters and 420 MHz in one package.

VHF-FM rigs come in three flavors: base, mobile and hand-held. Base rigs are often multimode, with SSB, CW and FM. Mobile radios are much smaller and generally offer FM only. Some mobile transceivers pack a wallop with 50-75 watts output, although they often include a low-power setting of 5 watts or less. FCC rules and good amateur practice obligate us to use the minimum power necessary to maintain communications.

Hand-held transceivers have undergone the biggest change over the years. They've become incredibly small, with some models hardly bigger than a box of cough drops. Despite their size, many models produce more than 5 watts of RF output. Memories, digital readout, dual bands—today's hand-held rigs have it all!

Which rig is for you, multimode, mobile or hand-held? Because we're dealing with FM here, let's lay multimode rigs aside. That leaves a mobile or hand-held. If you're in town or close to your favorite repeater, a hand-held transceiver may be for you. On the other hand, mobile transceivers are ideal for the ham who spends a lot of time on the road.

A mobile transceiver can be permanently mounted in the dashboard of your vehicle. If you intend to remove your transceiver for use at home (or for protection against theft), a slide-out bracket beneath the dashboard is a convenient

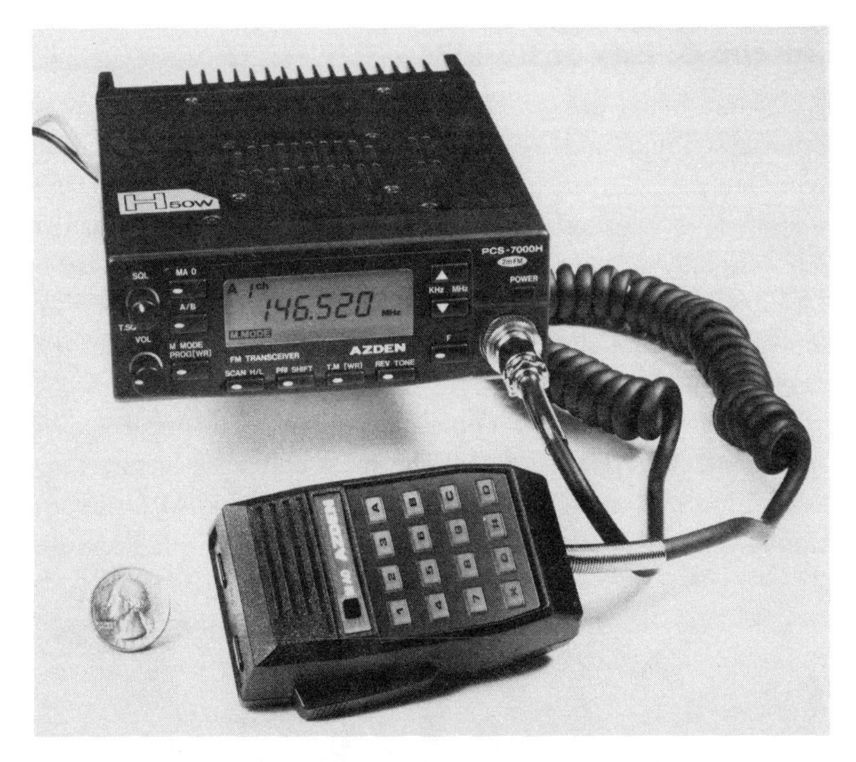

Modern FM transceivers have continued to shrink in size, while offering an ever-expanding selection of convenient features. The unit shown here is an Azden model PCS-7000H.

alternative. Regardless of which mounting configuration you choose, be sure to follow the electrical wiring guidelines in your manual.

Your hand-held transceiver can also function as a mobile rig. You can even power it from your vehicle's electrical system. (Voltage converters are necessary for some hand-helds. Consult your manual *before* attempting to connect the transceiver to the electrical system.) Many hams add amplifiers and outside antennas to increase the effective range of their mobile hand-helds.

Antennas: Buy or Build?

Most hams use ground-plane antennas for FM. They're omnidirectional and vertically polarized, ideal for repeater work and perfectly suited for communicating with mobiles. VHF's line-of-sight characteristic makes polarization more critical than on HF. Virtually all repeaters transmit vertically polarized signals. Using a horizontally polarized antenna sacrifices gain and limits the range of your station. Besides, ground planes are small and easy to build (see Fig 2-2).

What about beams? You haven't seen beams until you've seen some VHF/UHF arrays! The most elaborate ones are reserved for SSB and CW, but Yagis have been on FM for a long time—vertically polarized, of course. So has the quad and the

The Kenwood TH31BT (left) and the Yaesu FT-209H (right) are typical examples of FM hand-held transceivers. These hand-helds are complete portable stations, with autopatch keypads and flexible antennas.

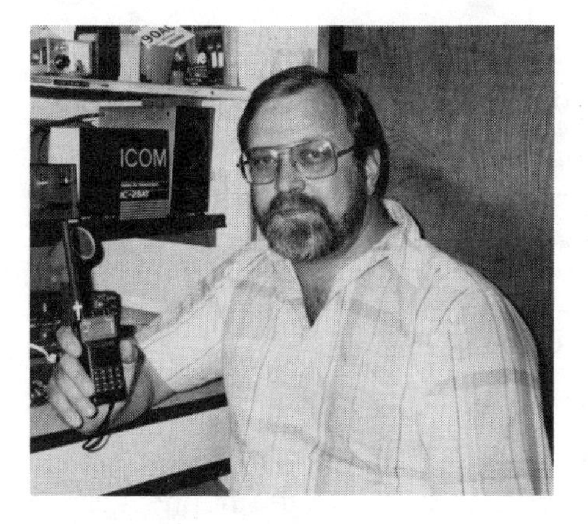

John Raifsnider, WA8OKA, shows off his pride and joy: a new ICOM IC-2SAT hand-held transceiver. Note the size of the rig. It's barely larger than John's hand!

Yagi/quad hybrid, the *quagi*. Many hams have built their own beams, but buying commercially made Yagis can save a lot of beginner headaches. Except in far-flung locations, beams aren't common on FM frequencies. They're really unnecessary if you let the repeaters do the work for you.

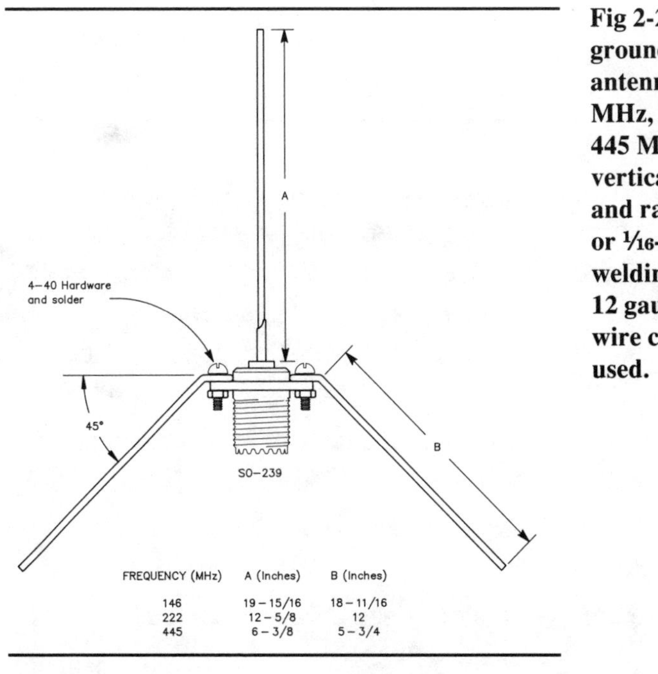

FREQUENCY (MHz)	A (Inches)	B (Inches)
146	19-15/16	18-11/16
222	12-5/8	12
445	6-3/8	5-3/4

Fig 2-2—A simple ground-plane antenna for 146 MHz, 222 MHz or 445 MHz. The vertical elements and radials are 3/32- or 1/16-in. brass welding rod. 10 or 12 gauge copper wire could also be used.

For mobile applications, you can build your own antenna, or buy any of the dozens of commercially manufactured models. You can mount mobile antennas on your trunk, on the roof (either permanently or with a magnetic mount) or on the glass of your rear window.

Make New Friends on FM

If you don't have friends on the air already, you'll make new friends soon enough. Attend club meetings and meet your fellow operators in person. If your club is assisting in a public service event, volunteer to help. By working closely with your new-found friends, you'll discover that the bonds between you will grow even stronger.

In time, you won't be "that new guy on the repeater." Everyone will know you and the conversations will flow easily—as they always do when friends are having fun together!

VHF-FM Glossary

access code: one or more numbers and/or symbols that are keyed into the repeater with a telephone tone pad to activate a repeater function, such as an autopatch.

autopatch: a device that interfaces a repeater to the telephone system to permit repeater users to make local telephone calls. Often just called a "patch."

break: the word used to interrupt a conversation on a repeater only to indicate that there is an emergency.

channel: the pair of frequencies (input and output) used by a repeater.

closed repeater: a repeater whose access is limited to a select group.

control operator: the Amateur Radio operator who is designated to "control" the operation of the repeater, as required by FCC regulations.

courtesy beep: an audible indication that a repeater user may go ahead and transmit.

coverage: the geographic area within which the repeater provides communications.

duplex: a mode of communication in which you transmit on one frequency and receive on another.

frequency coordinator: an individual or group responsible for assigning channels to new repeaters so they won't interfere with existing repeaters.

full quieting: a received signal that contains absolutely no noise.

hand-held: a portable transceiver small enough to fit in the palm of your hand or clipped to your belt. Sometimes called an HT or handie-talkie.

input frequency: the frequency of the repeater's receiver (and the output frequency of your transmitter).

key up: to turn on a repeater by transmitting on its input frequency.

machine: a repeater system (slang).

magnetic mount, mag-mount: a mobile antenna with a magnetic base that permits quick installation and removal from a motor vehicle.

NiCd: a nickel-cadmium battery that may be recharged many times; often used to power portable transceivers. Pronounced "NYE-cad."

open repeater: a repeater whose access is not limited.

continued on page 2-24

continued from page 2-23

output frequency: the frequency of the repeater's transmitter (and your transceiver's receiver).

simplex: a mode of communication in which you transmit and receive on the same frequency.

time-out: to cause the repeater or a repeater function to turn off because you have transmitted for too long.

timer: a device that measures the length of each transmission and causes the repeater or a repeater function to turn off after a transmission has exceeded a certain length.

tone pad: an array of 12 or 16 numbered keys that generate the standard telephone dual-tone multifrequency (DTMF) dialing signals.

CHAPTER 3

Are You Connected to Packet?

By Steve Ford, WB8IMY

I've reached the end of another long day. Now it's time to slip into casual clothes and get in touch with the world—the Amateur Radio world, that is.

I switch on my computer as I stroll into the shack. It rewards

A computer, a TNC and a 2-meter FM transceiver—that's all you need to become packet active!

me with a couple of beeps followed by the gentle whine of the cooling fan. While waiting for the machine to boot up, I turn on my trusty power supply, awakening the 2-meter FM transceiver and TNC. I twist the transceiver's VFO until 145.05 MHz appears on the digital display.

Within a few minutes, the computer presents me with its menu screen. I select **Packet** and wait for the hard disk to find my terminal program. Within seconds, my computer and TNC are communicating, displaying sporadic bursts of activity on the frequency.

Soon the local packet bulletin board (*PBBS*) sends its identification and "Mail for:" message. As I scan the list of call signs, I'm pleasantly surprised to see my own. I connect to the PBBS and pick up messages from friends in my home town and elsewhere. I type a few brief replies and send a *LIST* command to see any new bulletins that may have arrived. It looks like there are new orbital elements from AMSAT. I read the file and save it to disk.

After disconnecting from the PBBS, I reach for the transceiver's VFO. Maybe I should slide over to the PacketCluster frequency and see if there's any activity on 6 meters this afternoon. While I'm there I can get the latest solar conditions and check the QSL manager database for that 9L1 I worked last weekend—

The rhythmic chime of the *connect alarm* shatters my concentration. A message flashes across my monitor screen.

Hi Steve! Noticed you were on the air and wanted to know if you could help me with the club meeting tonight. I need someone to give a talk on satellites. What say? >>

Oh, sure! Whip together an hour-long presentation on satellites while I'm eating dinner. Is he serious? I begin to type, "Sorry, but I already have a commitment..."

Aw, what the heck. The only commitment I have is to my easy

chair. I hammer the BACKSPACE key several times and start over.

Satellites, eh? I suppose I can do that. When do you want the presentation to start? >>

Sometimes packet radio can be a little too convenient!

What is Packet Radio?

There is a long answer and a short answer to that question. The long answer goes beyond the scope of this book. Besides, there are several other books overflowing with detailed information on packet radio. (See the Resources and References Guide for more information.)

The short answer cuts right to the heart of this fascinating medium—and it's much easier to understand! Packet radio is computer-to-computer communication using radio links rather than telephone lines. Rather than sending a continuous stream of information, the data is assembled into neat little bundles called *packets*. These packets are sent one at a time. When a packet arrives error-free, the receiving station sends an "okay" message (called an *ACK*) and the next packet is transmitted. If the receiving station detects an error, it discards the faulty packet and does nothing. After a certain amount of time has passed without an ACK from the receiving station, the transmitting station sends the packet again. These exchanges take place at high speeds, resulting in efficient digital communications *without errors*—a remarkable feat for a radio-based computer network.

Amateur packet radio has already evolved beyond the dreams of its pioneers. Packet exists as a loose global network that is becoming more efficient every day. Through packet radio, you can tap into this continuously flowing stream of data and exchange information with stations throughout the nation and the world. From the fascinating to the absurd, it's all on packet!

Most VHF packet activity takes place on 2-meter FM. You'll

also find packet on 222 MHz and 420 MHz, although most of the activity on these bands is confined to *backbone* links that transfer packet data from one PBBS to another. (Individual users are discouraged—often prevented—from attempting to access backbone systems.) Packet is alive on 6 meters as well—with many opportunities for packet DX! See Table 3-1 for a list of popular packet frequencies.

What Can I Do With Packet?

The answer depends on your interests! Every amateur has different uses for packet. Some view packet as a tool to enhance their overall enjoyment of Amateur Radio. Others take it further, using packet as their primary means of communication.

Table 3-1

Popular Packet Frequencies

VHF

6 meters

Look for activity from 50.60 through 51.78 MHz.
50.62 is the 6-meter packet calling frequency.

2 meters

This the most popular VHF packet band. You'll find activity at 144.91, 144.93, 144.95, 144.97, 144.99, 145.01, 145.03, 145.05, 145.07 and 145.09 MHz. Packet can also be found between 145.50 and 145.80 MHz.

222 MHz and Up

Packet can be found on the 222, 420, 902 and 1240-MHz bands, but bulletin boards and live QSOs are somewhat sporadic. Most of the activity on the higher bands is in the form of *backbone* networks that pass packet traffic between bulletin boards and nodes. Avoid these backbones; they're not intended for individual user access.

To make things simple, let's list the most popular applications of VHF packet as it exists today:

- ❑ Enjoying live keyboard-to-keyboard QSOs.
- ❑ Accessing bulletin boards to exchange messages with other amateurs and read general interest bulletins. Some packet bulletin boards offer additional services such as electronic call sign directories and magazine bibliographies.
- ❑ Participating in the ARRL National Traffic System. (Packet bulletin boards can be used to receive and originate NTS traffic.)
- ❑ Using DX *PacketClusters* to hunt HF or VHF DX. Through *PacketClusters* you can determine which DX stations are on the air at the moment—and where they are! DX *PacketClusters* support other useful features as well.
- ❑ Monitoring and communicating with amateur satellites. Many satellites function as orbiting bulletin boards, relaying packet messages around the world. Others transmit images that you can display on your computer screen.

Before you can do any of these marvelous things, however, you need to have the proper equipment!

Assembling a Packet Station

A basic packet station consists of a 2-meter FM transceiver, a computer and a terminal node controller (TNC) or multimode communications processor (MCP). For 2-meter packet operating, fancy antennas are *not* required. A simple ground plane is fine for most areas. If packet activity is hot and heavy where you live, just the rubber duck antenna on a hand-held transceiver may be adequate.

The Computer

Just about any type of computer will be perfect for packet if

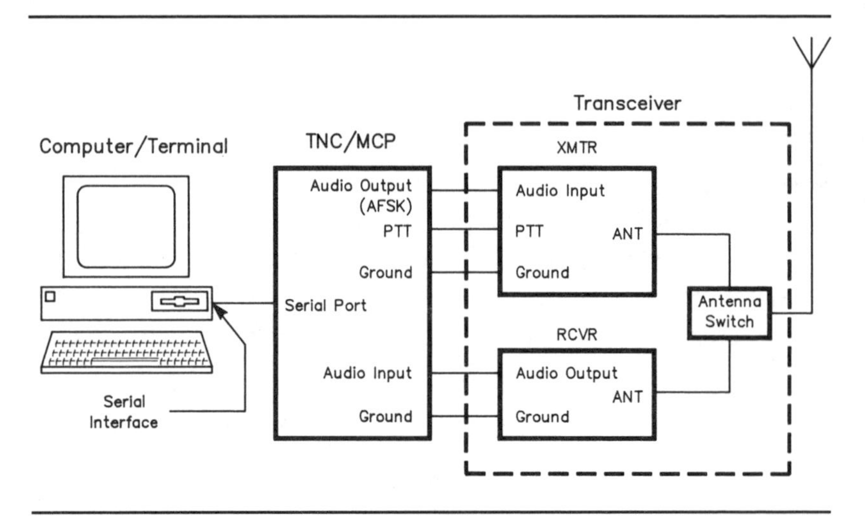

Fig 3-1—A diagram of a typical packet station.

it meets the following requirements:

- ❏ Provides an EIA-232-D (RS-232-C) or TTL serial port. This is the pathway the computer will use to communicate with the TNC or MCP and vice versa. (Some TNCs are designed for installation *inside* IBM PC-type computers and do not require a separate computer communications port.)
- ❏ Is able to run *terminal emulation* software. A terminal is a simple keyboard and monitor package that's used to communicate with computer systems. Since your TNC/MCP contains a microprocessor computer, it also requires a terminal to communicate with its human operator. By running a terminal emulation program, your computer will behave just like a terminal as far as the TNC or MCP is concerned. Table 3-2 lists packet software packages for a variety of computers. If you already own software for communicating with telephone data bases (such as CompuServe), chances are the software will also be usable for packet.

If you have a computer gathering dust in your closet, drag

Multimode communications processors such as these offer packet and several other digital modes.

it out and put it on packet! You can also find Commodores, Tandy Color Computers, IBM PCs and other computers selling for bargain prices at hamfest flea markets. Even a basic data terminal unit—often available at flea markets for $10 or $20—will do the job.

The TNC/MCP

A terminal node controller or a multimode communications

Multimode communications processors incorporating digital signal processing (such as this AEA DSP-2232) offer extraordinary flexibility. Depending on the internal software, a DSP processor can communicate in almost any data format!

processor forms the heart of any packet station. What's the difference between a TNC and an MCP? When it comes to packet, there really is no difference. If you want to operate other digital modes, however, the MCP is more versatile. An MCP offers RTTY, AMTOR, CW and other modes *in addition* to packet. A

Table 3-2

A Packet Software Sampler

Please note that the products and addresses shown below are subject to change. Contact the distributors to verify availability before ordering.

Apple II

APR: Send a blank, formatted diskette and a postage-paid, self addressed disk mailer to Larry East, W1HUE, PO Box 51445, Idaho Falls, ID 83405-1445

Apple Macintosh

MacRATT: Advanced Electronic Applications (AEA), PO Box C-2160, Lynnwood, WA 98036-0918

Atari

PK2: Electrosoft, 3413 N Duffield Ave, Loveland, CO 80538

Commodore

TNC64: Texas Packet Radio Society, PO Box 50238, Denton, TX 76206-0238

DIGICOM>64: A & A Engineering, 2521 West LaPalma, Suite K, Anaheim, CA 92801 tel 714-952-2114

MFJ Enterprises, Box 494, Mississippi State, MS 39762 tel 601-323-5869

(*DIGICOM>64* is a TNC *emulation* program that causes the computer to behave as a TNC. The only external device required is an inexpensive modem interface. The software and modem can be purchased separately.)

TNC is a packet-only device.

The choice to buy one or the other depends on your goals. Do you want to try VHF RTTY or CW? Does your license class include HF privileges or do you plan to upgrade soon? If so, an MCP may be the better choice. On the other hand, if you are

---

IBM PC

LAN-LINK: Joe Kasser, G3ZCZ, PO Box 3419, Silver Spring, MD 20918. Evaluation copy available for $5.

PK GOLD Enhanced: InterFlex Systems Design Corp, PO Box 6418, Laguna Niguel, CA 92607-6418.

PC-Pakratt II: Advanced Electronic Applications (AEA), PO Box C-2160, Lynnwood, WA 98036-0918.

BayCom: PacComm Inc, 4413 N Hesperides St, Tampa, FL 33614-7618, tel 813-874-2980

Poor Man's Packet (PMP): For information, send a self-addressed, stamped envelope to F. Kevin Feeney, WB2EMS, PO Box 323, Newfield, NY 14867.

(*BayCom* and *PMP* are terminal emulation programs similar to *DIGICOM>64)*

Tandy Color Computer

COCOPACT: For more information, send a self-addressed, stamped envelope to Monty Haley, WJ5W, Route 1, Box 150-A, Evening Shade, AR 72532

Texas Instruments TI-99/4A

Mass Transfer: A general-purpose terminal program (on disk). Send a self-addressed, stamped envelope to Stuart Olson, 6625 W Coolidge St, Phoenix, AZ 85033

interested only in packet, save your money and buy a basic TNC.

The primary function of a TNC/MCP is to receive digital data from your computer and turn it into properly assembled packets. This assembly process includes the translation of digital data into audio tones for transmission. Conversely, a TNC/MCP disassembles received packets, translating audio from your transceiver into digital data your computer will understand. Other important functions include acknowledging error-free packets, or retransmitting unacknowledged packets.

TNCs and MCPs are *smart* devices, containing their own microprocessors, memory and software. They respond to specialized commands from your computer and, through the use of these commands, can be tailored to match your particular operating habits. Your TNC/MCP manual will describe all available commands in detail.

It is also possible for some computers to function as TNCs through the use of specialized software. All that's required is a simple outboard modem circuit to act as the interface between the computer and your radio. *DIGICOM>64* is a software-based TNC system for Commodore computers. *BayCom* and *Poor Man's Packet* (or *PMP*) are designed for IBM PCs and compatibles. These shortcut systems are not as versatile as standard TNCs, but they'll get you on the air quickly and inexpensively (see Table 3-2).

The Radio

Almost all FM transceivers will serve as packet radios. The only exceptions are rigs that can't switch from transmit to receive quickly, or rigs that distort the packet tones during transmission or reception. Fortunately, these problems are not common.

Connecting your transceiver to your TNC or MCP is a matter of soldering the correct wires to the correct pins. Most TNC/MCP manufacturers provide a cable for your transceiver, but they usually *do not* provide the microphone connector. If you don't have a spare connector in your junk box, you'll need to contact the transceiver manufacturer or make a trip to your local parts shop.

Your TNC/MCP manual will tell you which wire in the cable is used for which purpose. When assembling your microphone connector, there are two wires of greatest concern: *AFSK* (audio frequency shift keying) and *PTT* (push to talk). AFSK refers to the packet tones that are transmitted when the radio is keyed. The AFSK wire carries these tones and must be attached to the *audio input* pin of the microphone connector. To enable the TNC/MCP to key the transceiver automatically, you have to connect the PTT (push to talk) leads to the PTT pins. Wiring the microphone connector can be tricky with hand-held transceivers. Fig 3-2 suggests some ideas you may wish to try.

Use great care when wiring your microphone connector. *Don't plug it in* until you've checked your finished handiwork thoroughly. If you're uncertain about how to wire the connector, consult your transceiver manual or contact the manufacturer. Some radios, for example, provide 12 volts at the microphone jack to power external devices such as tone (DTMF) pads. If you connect one of your TNC/MCP wires to that particular pin, the results could be disastrous!

Once you have the microphone connector wired properly, the rest is easy. The receive audio for the TNC/MCP is usually obtained at the external speaker or earphone jack. Many TNC/MCP manufacturers provide a cable with a pre-wired ⅛-inch diameter plug. If this plug doesn't fit your radio, you'll need to replace it with another of the correct size or use an adapter. If your radio doesn't have an external speaker or earphone jack at all, you'll need to use a Y connector to tap the audio at the speaker.

With luck, you'll be on the air from the moment you attach your connectors and power-up your equipment. On the other

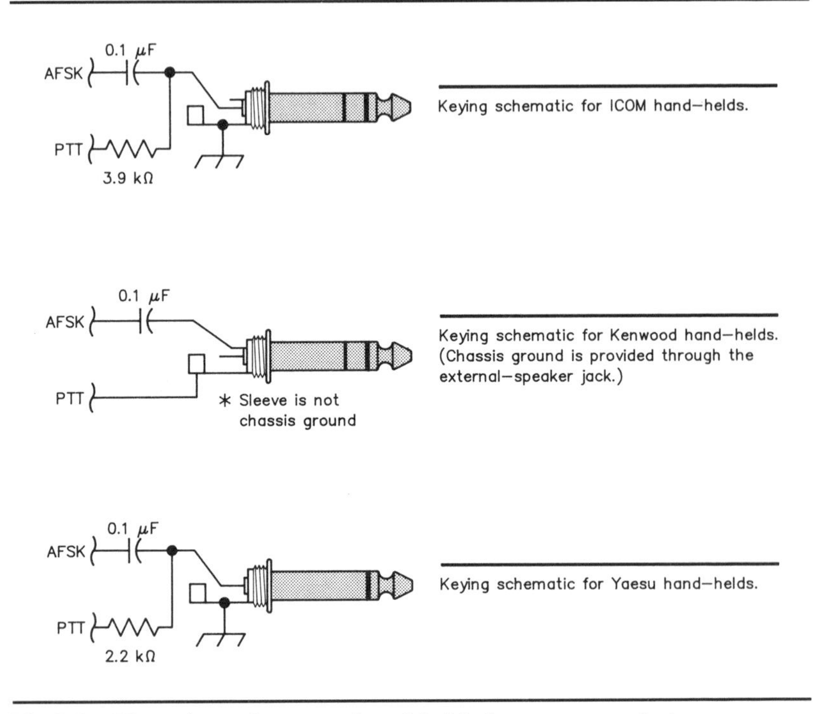

Fig 3-2—If you intend to use a hand-held rig as your packet transceiver, here are some suggestions for microphone and push-to-talk configurations. (Component values are approximate.) When in doubt, however, consult your manuals.

hand, you may discover that some adjustments are necessary.

You may overdrive or underdrive your microphone input, for example. TNCs and MCPs provide level-adjustment procedures to solve this problem. If you can't seem to copy signals that you *know* are present, check the VOLUME control on your transceiver. You may have it too high or too low. Setting it at the ten o'clock position is usually sufficient.

Check your SQUELCH control as well. Too little squelch will allow weak signals and noise to "trick" your TNC/MCP into thinking that a signal is present. (It won't key your rig until it detects a period of silence.) Too much squelch may block reception of all packet signals!

Finally, wrap your cables neatly and keep them away from the coax or, in the case of hand-held rigs, the antenna. RF can sneak into TNC/MCP cables and drive the units crazy. With all your wiring accomplished, it's time to start communicating!

Live Packet QSOs

The easiest way to get started in packet is to make a *connection* to another station and have a live (real-time) QSO. A TNC/MCP is connected to another station when it exchanges the initial protocols and begins the error-checking routine described earlier. If you have a friend active on packet, make an appointment to get on the air together and have a chat. Otherwise, monitor a frequency and watch for the call signs of stations you may wish to contact.

To start the QSO, you have to issue a *connect request*. In simple terms, your TNC or MCP will attempt to contact the other station and "request" a connection. Use CONTROL-C (or whatever key combinations your software requires) to place your TNC or MCP in the *command* mode. Everything you enter from the keyboard will now be interpreted as a command to be executed. You don't have to enter the full command. Just an abbreviation will do in most cases. For the sake of clarity, however, I'm going to show the complete command. Type carefully! If the TNC or MCP doesn't understand your request, you'll see "EH?", "HUH?" or a similar response!

cmd: CONNECT N6ATQ <CR>

(Throughout this chapter we'll use <CR> to represent the ENTER or RETURN key on your keyboard.)

You've just asked your TNC/MCP to contact N6ATQ. Watch your transceiver. It should begin transmitting short packet bursts automatically. (If another station is transmitting, your rig will not be keyed until the other transmission stops. This technique allows several packet stations to share the same frequency.) When the connection is finally established, you'll see the following:

***** CONNECTED to N6ATQ**

Congratulations! You're officially packet-active! You'll notice that the cmd: prompt has disappeared. This means that your TNC/MCP has left the command mode and has entered the *converse* mode. Everything you type will now be interpreted as text to send to N6ATQ rather than commands for your TNC/MCP.

The rest is up to you. The QSO proceeds like any other, except that you're typing rather than talking or tapping a key. It's important to let the other station know when you're finished with a thought and awaiting his or her response. Without some method to control the "flow" of the conversation, a packet QSO can get pretty confusing! Most packeteers use "K" or ">>" to mean "over." For example:

Hello, Craig. My name is John and I live in Oceanside >> <CR>

Hi, John! Nice signal here in Escondido. How long have you been on packet? >> <CR>

If you have an outside antenna and a reasonable amount of output power (20 or 30 watts), you can enjoy a number of direct, point-to-point QSOs. But what if you attempt to make a connection and the other station can't hear you? Your TNC/MCP will try to establish communications, but after a certain number of attempts (usually 10) you'll see:

Retry count exceeded
***** DISCONNECTED**

This is the TNC's way of saying, "I give up!" It looks like you need an intermediate station to act as a relay, doesn't it?

That's where *nodes* and *digipeaters* come in!

Packet Nodes and Digipeaters

Many hams who are new to packet confuse nodes and digipeaters with voice repeaters. There are similarities, but there are major differences as well...

- ❏ Voice repeaters listen on one frequency and transmit on another. Most nodes and digipeaters listen and transmit on the *same* frequency. The exceptions are nodes that act as links to other stations via UHF or VHF backbone systems, or nodes that function as *gateways* to HF packet frequencies.
- ❏ Voice repeaters relay *every* signal heard on the input frequency. Nodes and digipeaters relay only packets addressed specifically to them.
- ❏ Voice repeaters are *real-time* devices. They listen and transmit simultaneously. Nodes and digipeaters receive packets, store them momentarily and then retransmit.
- ❏ Only one station at a time can use a voice repeater. Several packet stations can use a node simultaneously.

In the early days of amateur packet, nodes didn't exist. If you needed a relay you had to use a digipeater. Every second-generation TNC (called a *TNC2*) had the ability to function as a digipeater, so any station could act as a relay for another. (This is still true today, although it's not a common practice.) Some amateurs established powerful digipeaters by placing TNCs, transceivers and antennas at excellent operating sites. This allowed hams in poor locations to extend their coverage over wide areas.

The big problem with digipeaters was that they weren't "intelligent." They shuffled data from one station to another and nothing else. If you connected to a station through several digipeaters, every bit of information had to travel from one end

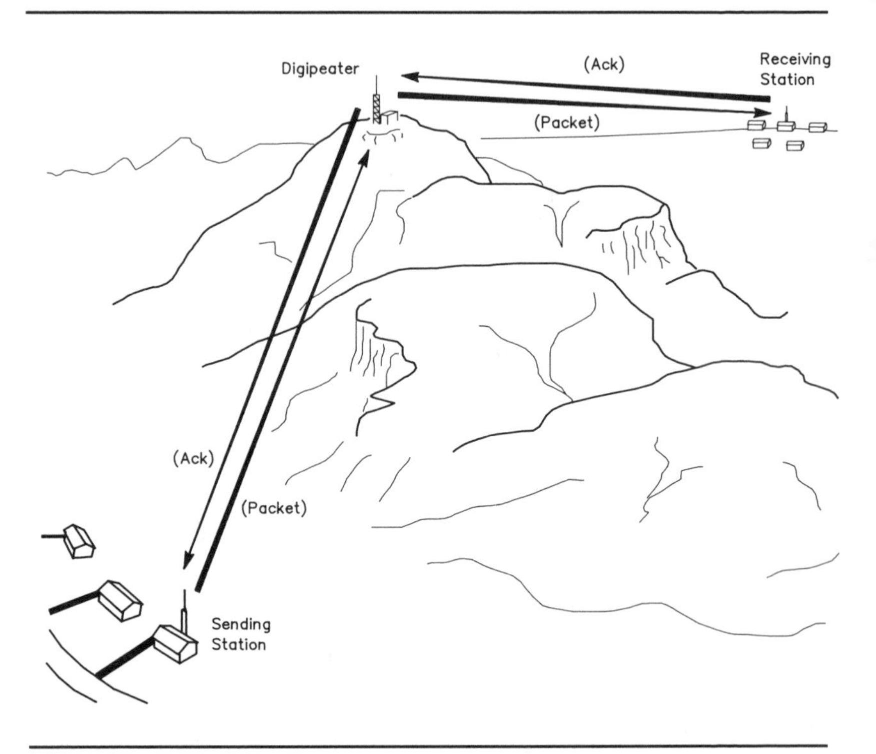

Fig 3-3—In the early days of packet, digipeaters spanned the gaps between distant stations. However, every bit of packet data had to travel back and forth through the digipeater from one station to the other. In this example, the packet and the ACK must complete the entire journey from end to end.

of the path to the other (see Fig 3-3). The more digipeaters involved in the path, the greater the possibility of failure.

When the first *NET/ROM* and *TheNet* nodes hit the packet scene, the change was startling. All you had to do was get the packet to the node and the node took responsibility for relaying the packet to its destination. For example, if the receiving station received the packet error-free, its ACK only had to travel to the node—not all the way back to your station (see Fig 3-4)! By eliminating end-to-end acknowledgments, efficiency improved dramatically.

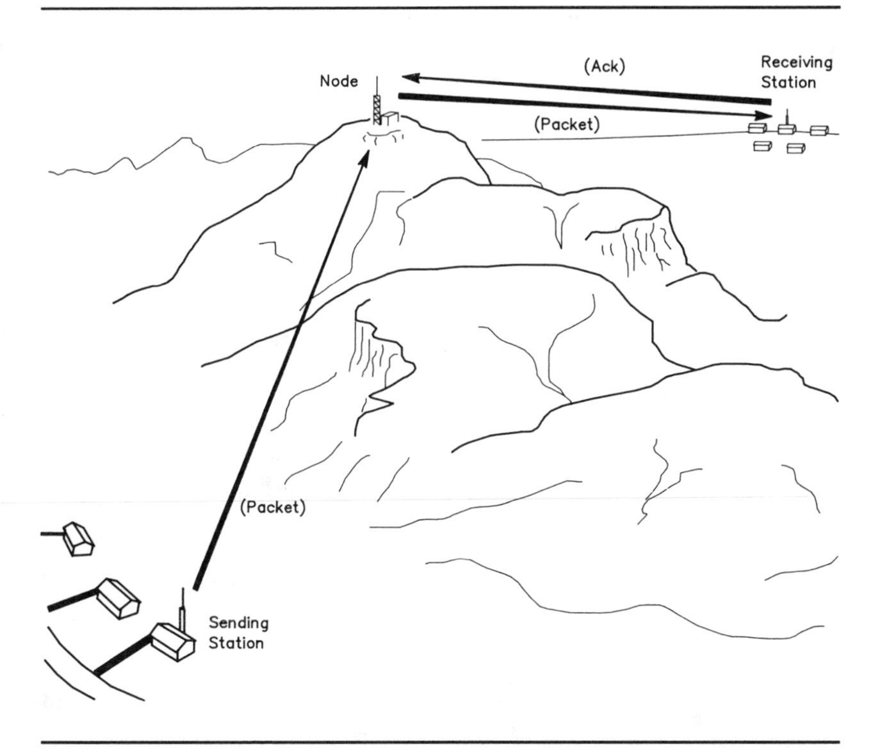

Fig 3-4—Packet nodes are more efficient than digipeaters. The node relays the data in this example, but the ACK from the receiving station only needs to travel back to the node.

In our previous example, you could have used a node to make the connection to N6ATQ. Let's call our node WA6ZWJ-5. You begin by establishing a connection to the node.

cmd:CONNECT WA6ZWJ-5 <CR>

If the node acknowledges your request, you'll see:

***** CONNECTED to WA6ZWJ-5**

Now you can use the node to reach N6ATQ. Simply enter:

CONNECT N6ATQ <CR>

You've asked the node to contact N6ATQ for you. With any luck you'll be rewarded with:

***** CONNECTED to N6ATQ**

By the way, you're probably wondering why the call sign of the node in our example has a "-5" following the last letter. That's called a *secondary station identifier*, or *SSID*. It's used to separate one function from another at the same station location. WA6ZWJ might also have a bulletin board, for example. If so, he may choose to call it WA6ZWJ-**4**. By using the full call sign and the proper SSID, you can connect to the station you desire (the node or PBBS, in this case).

In addition to their call signs, packet nodes can also be identified by their *aliases*. Aliases come in handy when you can't seem to remember the node you want to reach. They often indicate the location of the node or the name of the group that operates it. For example, the alias of WA6ZWJ-5 might be RIVER—an abbreviation for Riverside. You can command your TNC/MCP to connect to WA6ZWJ-5 or RIVER. The connection will be made regardless of whether you use the call sign or the alias.

If the distance between you and another station is beyond the capability of one node to handle alone, you can link several nodes together to form a multi-node path. All you need to know is the call sign of the node nearest to the station you want to contact. (Let's call that node WB6YUE-5.) Once you have this bit of information, connect to your local node and "ask" it to make the connection to the destination node.

CONNECT WB6YUE-5 <CR>

Now just sit back and relax. The network will take care of everything—even using 222- or 420-MHz backbones if necessary to reach the destination! Each node maintains a list of other nodes it can reach reliably. This list is updated frequently. If the network is able to establish a path to WB6YUE-5, you'll see a message similar to the one shown below.

WA6ZWJ-5> Connected to WB6YUE-5

All you have to do is issue your final connect request to the station you wish to contact:

CONNECT N6ATQ <CR>

It's important to point out that QSOs over long paths (linking more than two or three nodes) are discouraged because they tend to reduce the efficiency of the network. Longer paths are also unreliable and prone to failure. Use your best judgment before attempting to communicate in this fashion.

With node networks springing up throughout the country, packet operating has become easier than ever before. Digipeaters still exist, but nodes now dominate most packet networks. Unless you live in a remote area, there is probably a node you can use to connect to many other stations—such as bulletin boards!

Packet Bulletin Boards

Bulletin boards (PBBSs) form the hubs of most VHF packet networks. They're electronic warehouses for the bulletins, private mail and NTS traffic that flow through the packet system. By connecting to a PBBS, you'll be able to read general bulletins, send mail to other packet-active hams or send traffic to just about anyone!

Most PBBSs are operated by clubs or private individuals. It takes a fair amount of time and money to maintain such a system. The system operator—or SysOp—is the person who calls the shots on any PBBS. After all, it's his time and, in many cases, his money! The worldwide packet network depends upon the dedication of SysOps. When you consider that this is an all-volunteer effort, it is amazing that the system functions as well as it does.

Connecting to a Bulletin Board

Connecting to a PBBS is the same as connecting to any other packet station. Monitor the packet frequencies in your area and watch for bulletin board activity. When you find one, make your connect request and have fun! Let's pretend we've discovered the W1NRG-4 PBBS.

cmd:CONNECT W1NRG-4 <CR>

We'll also presume that you're close enough to connect directly. If not, you'd use a node to bridge the gap.

***** CONNECTED TO W1NRG-4**

Good! You've made the connection. Now the PBBS will send its sign-on information to you.

Welcome to W1NRG's MSYS PBBS in Wallingford, CT
Enter command: A,B,C,D,G,H,I,J,K,L,M,N,P,R,S,T,U,V,W,X,?,* >

That's a big list of commands and we don't have the space to discuss each one! Let's try the most popular functions instead. Would you like to see some of the bulletins available on the PBBS today? The *L*, or *LIST*, command will do the trick!

MSG #	TR	SIZE	TO	FROM	@BBS	DATE	TITLE
744	B#	473	WANTED	K1CC	USBBS	920520	AT motherboard
743	B#	654	ALL	N1GFL	NEBBS	920520	HF RIG FOR SALE
742	N	875	WA1TRY	KA1TMN	WA1TRY	920520	LAST PROBLEM
739	B#	652	NEEDED	KO1C	USBBS	920519	DENTRON MANUAL
736	B#	488	WANT	KA1TYV	USBBS	920519	SOFTWARE HP 100/150
735	TF	424	N1EOA	WA1MVJ	K1UGM	920519	IMPORTANT
734	TF	1230	N1EOA	WA1MVJ	K1UGM	920519	FT470 MOD VHF
733	B#	853	INFO	N1IPO	USBBS	920519	MY CALL SIGN?
730	B#	949	HELP	WA3YHH	USBBS	920519	MODS TO IC-22A
710	B#	5872	SWL	AE1T	NEBBS	920519	English Broadcasts
708	B#	611	ALL	W1OER	NEBBS	920519	FOR SALE:ICOM 228H

You can read any of these messages by sending the letter *R* followed by the message number. Sending "R 743" allows you to read message number 743 concerning an HF rig for sale.

Sending Packet Mail

Would you like to send a message to a packet-active amateur in another city, state or country? If you know the call sign of his local PBBS, it's easy! Just enter:

SP N6ATQ @ WA6ZWJ <CR>

This odd-looking line translates to: Send a private message (SP) to N6ATQ at (@) the WA6ZWJ bulletin board. The PBBS now asks for the subject of the message:

SUBJECT?:

Enter a brief subject sentence. Keep it *very* short.

WORKED MEXICO ON 6 METERS! <CR>

The PBBS will respond with something like this:

ENTER MESSAGE. USE CTRL-Z OR /EX TO END MESSAGE.

That's your cue to begin entering the text of your message. If the message is going to travel out of your region, keep it as brief as possible. Longer messages travel much slower. When you've finished entering your message, enter CONTROL-Z or /EX on a line by itself. You'll soon see the PBBS *command menu* again.

Enter command: A,B,C,D,G,H,I,J,K,L,M,N,P,R,S,T,U,V,W,X,?,* >

How long will it take for the message to reach your friend? It depends on a host of factors, including propagation. The packet network is not a commercial system. You can't expect the same level of reliability. Your message may arrive in a few hours, a few days, a few weeks or not at all! Even so, packet mail is surprisingly reliable and efficient. I can usually send messages from Connecticut to Ohio, for example, in about 12 hours.

Sending a Bulletin

Sending a bulletin is similar to sending private mail. The only difference is that you don't have a particular person or destination in mind. Let's say you need help with a problem...

SB ALL @ USBBS <CR>

This means: Send a bulletin (SB) to everyone (ALL) at (@) every PBBS in the US (USBBS). This doesn't guarantee that everyone will see your message, but many will!

SUBJECT?

Describe your problem in as few words as possible.

HELP! ANTENNA ROTATOR STUCK! <CR>

As before, enter your text and end your message. Connect to the PBBS a few days later and you may have some helpful mail waiting! You can also use a packet bulletin to announce that you have an item for sale. However, FCC rules specify that the item must be directly related to Amateur Radio activity (a transceiver, an antenna, a computer, a monitor and so on). Don't try to sell your car or your house on the packet network!

Sending NTS Traffic

NTS traffic can be originated and received at most packet bulletin boards. Not only is this useful in emergencies, it's also handy for communicating with nonhams. Let's say I want to send an NTS message to my parents. I'd begin by entering:

ST 45429 @ NTSOH <CR>

This means: Send traffic (ST) to my parents' ZIP code (45429) at (@) National Traffic System/Ohio (NTSOH). Standard two-letter state postal abbreviations are used. If my parents were in California, I would send it to NTSCA.

When the PBBS asks for a subject, I respond with:

QTC Kettering (513) 555 <CR>

This cryptic line simply translates to: Traffic for (QTC) the city of Kettering with the telephone area code and prefix as shown. Now I enter my message in NTS packet format. The first line contains information about the nature of the message:

NR 3, ROUTINE, WB8IMY, 26, WALLINGFORD CT, DEC 18 <CR>

Can you guess what this means? It's not as complicated as it looks: Message number 3, routine priority, from WB8IMY, with a total of 26 words, originating from Wallingford, Connecticut, on December 18. Now I have to supply the name, address and complete telephone number of the recipient.

Mr & Mrs Robert Ford <CR>
549 Shady Lane <CR>
Kettering, OH 45429 <CR>
(513) 555-3958 <CR>

Finally, I enter my message:

Sending you a special Christmas surprise. Should arrive via UPS on Wednesday or Thursday. Hope you enjoy it. Will call Christmas day. Love, Steve and Kathy. <CR>

/EX <CR>

That's all there is to it! The packet bulletin board will relay the message to the network as soon as possible. At some point it may be intercepted and relayed on CW or phone nets. Either way, an amateur in the Kettering area will call my parents and deliver the happy message.

You can deliver traffic, too! Check your PBBS for traffic

using the *LT* (*List Traffic*) command. As you read the list of messages, do you see any intended for your area? If so, use the *R* command to read the traffic. Write the message down or print it on your printer. If you decide to deliver it, use the *KT* (*Kill Traffic*) command to delete the message from the PBBS. Delivering NTS traffic is a pleasurable experience and may turn you into a dedicated traffic handler!

These examples do not do justice to the full potential of PBBSs. Your best bet is to experiment and ask questions. Besides, PBBSs use different types of software. The commands and prompts are similar, but there are some important differences, too.

As you get to know your local system, you'll discover that it has a certain addictive quality! If you're curious about what's going on in the world of Amateur Radio, you'll feel compelled to connect to the PBBS every few days. As you make friends in the packet world, you'll also find yourself sending and receiving mail on a regular basis!

DX *PacketClusters*

Even if you lack privileges on HF, a DX *PacketCluster* has something to offer! Not all areas of the country are blessed with *PacketClusters*, but their population is growing rapidly. You may stumble upon one as you're monitoring your local packet frequencies. At first glance it may look like a bulletin board, but it isn't!

A *PacketCluster* is a network of nodes operating under specialized *PacketCluster* software. They're dedicated to contest and DX activities. You connect to a *PacketCluster* in the same way you'd connect to any other station. However, the information you'll receive will be very different!

cmd: CONNECT KC8PE <CR>
***** CONNECTED to KC8PE**

Welcome to YCCC PacketCluster node - Cheshire CT
Cluster: 22 nodes, 9 local / 144 total users Max users 367
WB8IMY de KC8PE 5-May-1992 0124Z Type H or ? for help >

The cluster is waiting for your command. What would you like to see? How about a list of the latest DX sightings (called *spots*)? Enter: **SHOW/DX <CR>**

7015.5	**UL7MG**	**5-May-1992**	**0119Z**	**<W3XU>**
14029.3	**UD6DFF**	**5-May-1992**	**0106Z**	**<K2LE>**
14211.9	**J68AJ**	**5-May-1992**	**0056Z**	**<NE3F>**
14002.6	**4S7WP**	**5-May-1992**	**0054Z**	**<K2LE>**
14007.0	**ZA1ED**	**5-May-1992**	**0048Z**	**<K2LE>**

WB8IMY de KC8PE 5-May 0124Z >

Now you have a list of the five most recent DX spots along with their frequencies and the times (in UTC) when they were heard. If you're tuning through the bands you may discover another DX station worthy of a spot on the cluster. Go ahead and make a contribution by posting it on the network. The simplest command format would be: **DX 21.250 SV3AQR <CR>**

If you don't have HF privileges, keep an eye on your local *PacketCluster* anyway. When significant VHF band openings occur, you'll often hear about it first on the *PacketCluster*!

PacketClusters have other useful features. You can use the **DIR** command to see a list of bulletins just as you would on a packet PBBS. Using the **R** (READ) command will allow you to read any bulletin you wish. Unlike packet bulletin boards, however, *PacketClusters* can relay bulletins and messages only within the networks they serve.

Packet and Satellites

There are several amateur satellites in orbit that you can access or monitor via packet (see Table 3-3). The only catch is that many of these satellites use signal formats (such as *Phase Shift Keying*, or *PSK*) that are not compatible with standard TNCs and MCPs. To use these satellites, you can add an external modem to your TNC to convert the signals to the proper format.

Satellite operating is fun and rewarding. Satellite bulletin boards give you the opportunity to exchange messages internationally with very little delay compared to terrestrial networks.

Table 3-3

Active Packet Radio Satellites

Satellite	*Packet Radio Mode*	*Uplink (MHz)*	*Downlink (MHz)*
PACSAT(AO-16)	1200-bits/s PSK AX.25	145.900/920/940/960	437.025 437.050(sec) 2401.100
DOVE(DO-17)	1200-bit/s AFSK	(Note 1)	145.825 2401.220
WEBERSAT(WO-18)	1200-bit/s PSK AX.25	(Note 2)	437.075 437.100(sec)
LUSAT(LO-19)	1200-bit/s PSK AX.25	145.900/880/860/840	437.150 437.125(sec)
FUJI(FO-20)	1200-bit/s PSK AX.25	145.850/870/890/910	435.910
RS-14(AO-21)	(Note 3)	435.016	145.983
	2400 bit/s BPSK Biphase-S	435.115	145.983
	4800 bit/s RSM NRZIC Biphase-M	435.193	145.983
	9600 bit/s RSM NRZI (NRZ-S) +Scrambler	435.193	145.983
UoSat-OSCAR 22	9600-bit/s FSK	145.900/145.975	435.120

Note 1. DOVE is not user-accessible; instead, it transmits digitized voice messages and AFSK AX.25 telemetry.

Note 2. WEBERSAT is not user-accessible; instead, it transmits Earth images using PSK AX.25 packet radio.

Note 3. Uplink uses 1200-bit/s FSK, NRZIC Biphase-M. Downlink uses 1200-bit/s BPSK, NRZI (NRZ-S).

(Source: *AMSAT-DL Journal*)

Large antennas and high power levels are not required to work these satellites. Most amateur packet satellites can be accessed with 20 watts and a ground-plane antenna!

If you want a taste of satellite packet activity without investing in extra hardware, try the DOVE satellite. You can't transmit to DOVE, but you can easily receive its packet telemetry using a standard TNC. A portion of an actual DOVE transmission is shown in Fig 3-5. The telemetry data is sent in *hexadecimal* format and there are programs available to decode the information into plain English. (See the Resources and References Guide for more information.)

You can also use your current packet equipment to contact the *mailbox* aboard the Russian *Mir* space station. Fig 3-6 shows actual packet transmissions from *Mir*. Like the DOVE satellite, *Mir* has a loud signal that's easily heard by the most modest ground stations. When *Mir* is within range, you send a connect request in the same manner as you would to your local PBBS. However, the mailbox is occasionally deactivated so that the cosmonauts can enjoy voice QSOs. To learn more about satellite

```
DOVE-1>TLM
21:CB 22:78 23:22 24:1B 25:1F 26:29 27:00 28:44 29:01 2A:02 2B:57
2C:00 2D:89 2E:67 2F:9E 30:CD 31:9B 32:00 33:01 34:A2 35:A0 36:A8
37:AC 38:A2

DOVE-1>STATUS
80 00 00 84 0A 18 FF 02 00 B0 00 00 0A 0C 3C 05 17 00 07 04 01

DOVE-1>BRAMST
NK6K and N4HY are putting together the voice software now. Stand by for
loading info. Telemetry still looks good.

DOVE-1>BCRXMT
vmax=759160 battop=766771 temp=357713

DOVE-1>LSTAT
I P:0x3000 o:0 l:14141 f:14141, d:0 st:0

DOVE-1>TIME-1
PHT: uptime is 014/21:49:53.
```

Fig 3-5—An actual transmission from the DOVE satellite. English text messages are often intermingled with telemetry data.

```
U5MIR-1>WB5KLY
Logged on to U5MIR's Personal Message System
CMD(B/H/J/K/KM/L/M/R/S/SR/V/?)>

U5MIR-1>WB5KLY
Subject?
```

Fig 3-6—The competition is fierce, but it's a real treat to connect with the packet mailbox on the Russian *Mir* space station!

communications—and how to track satellites in orbit—see Chapter 5: Satellite Communications.

The Future of Packet

Packet never stands still! Each year brings new improvements. Highly efficient network systems such as TCP/IP and ROSE are becoming popular. Innovative systems such as roundtable *chat* nodes and weather data nodes are appearing in many areas.

Ask yourself what *you* would like to do with packet radio. You can use any of the features we've already discussed, or create something special. How about a "private" packet link to a friend across town? You could exchange messages and software whenever the spirit moved you! Or maybe you'd like to put a PBBS on the air and become a SysOp yourself. Regardless of which activities you choose, if you want to be on the cutting edge of Amateur Radio technology, there's no better mode than packet!

CHAPTER 4

Beyond FM: CW/SSB Equipment and Operating

By Michael Owen, W9IP, and Rus Healy, NJ2L

FM operating on the VHF bands is fun, but there's even more excitement to be found on SSB and CW. Yes, SSB and CW operation is alive and well on VHF. In addition, you may be surprised to know that SSB and

Rick, NU7Z, enjoys VHF contest operating—not to mention a spectacular view—from his mountaintop location. If you love the great outdoors, a VHF contest offers the perfect opportunity to enjoy the best of both worlds!

CW equipment usually includes FM—and isn't much more expensive than FM-only gear.

Why bother with SSB and CW? Why not use only FM? For one thing, SSB and CW are more efficient modes than FM. In other words, the same power level can carry your signal farther on SSB and CW than FM.

You don't need repeaters to communicate over longer distances with SSB and CW. Except for the amateur satellites, all VHF SSB and CW operation is direct from station to station. FM operators refer to working without repeaters as *operating simplex*. SSB and CW operators wouldn't have it any other way! SSB and CW signals are detectable at levels where FM signals can't even be heard. For this reason, SSB and CW are called *weak-signal* modes. The signals aren't necessarily weak, but you can often copy them even when they are.

One benefit of weak-signal operation on VHF and UHF is the enormous amount of spectrum available. The 2-meter weak-signal band alone is wider than all the amateur bands from 1.8 to 30 MHz. Each band has nationally standardized *calling frequencies*. They save you time when looking for activity or favorable propagation (see Table 4-1).

The key to enjoyable use of this resource is to know how everyone else is using it, and to follow their lead. Essentially, this means to listen first. Pay attention to the segments of the band already in use, and follow the operating practices the experienced operators are using.

How Are The Bands Organized?

Weak-signal operating is more enjoyable when you know how the bands are organized. By knowing the best frequencies and times, you'll make plenty of contacts.

In most areas of the country, everyone uses the calling frequencies to establish contact. Then the two stations move up or down the band to chat. This way, everyone can share the calling frequency without having to listen to each other's QSOs. You can easily tell if the band is open by monitoring the call signs of the stations making contact on the calling frequency.

Weak-signal activity on VHF/UHF is concentrated on the two lower VHF bands, 6 and 2 meters (50 and 144 MHz). The number of active stations on these bands is about equal. Above the 2-meter band, there are considerably more active stations on the 70-cm (420 MHz) band than any other.

On 6 meters, a *DX window* has been established to reduce interference to DX stations. Yes, *DX* stations! During years of high solar activity, 6-meter openings to the other side of the world are possible! This window, which extends from 50.100-50.125 MHz, is intended for DX QSOs only. The DX calling frequency is 50.110 MHz. US and Canadian 6-meter operators should use the domestic calling frequency of 50.125 MHz for non-DX work. When contact is established, move off the calling frequency as quickly as possible.

Table 4-1

North American Calling Frequencies

Band (MHz)	*Calling Frequency*
50	50.110 DX 50.125 US, local
144	144.010 EME 144.100, 144.110 CW 144.200 SSB
222	222.100 CW/SSB
432	432.010 EME 432.100 CW/SSB

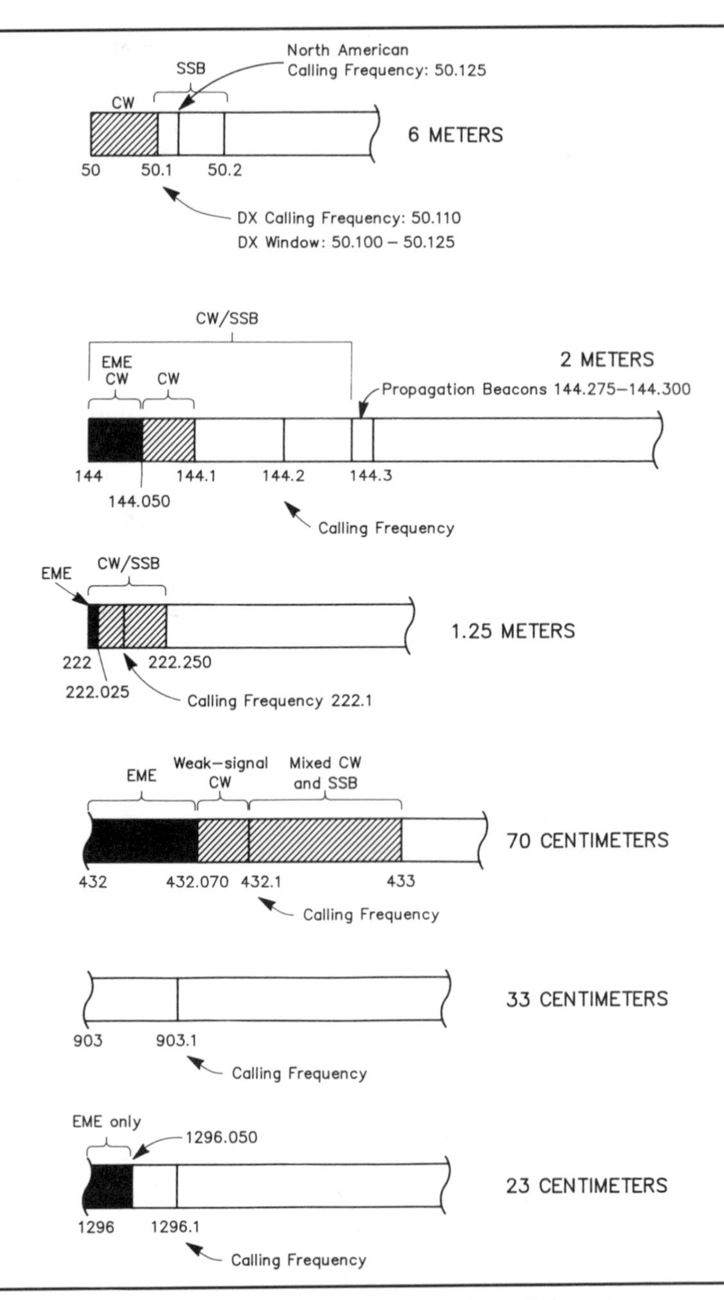

Fig 4-2—Suggested VHF/UHF band usage. On 222 MHz and above, activity is usually centered around calling frequencies.

Activity Nights

Although you can scare up a QSO on 50 or 144 MHz almost any evening (especially during the summer), in some areas of the country there isn't always enough activity to make it easy to find someone. Therefore, informal *activity nights* have been established. There's a lot of variation in activity nights from place to place. Check with an active VHFer near you to find out about local activity nights.

Common Activity Nights

Band (MHz)	*Day*	*Local Time*
50	Sunday	6:00 PM
144	Monday	7:00 PM
222	Tuesday	8:00 PM
432	Wednesday	9:00 PM
902	Friday	9:00 PM
1296 and up	Thursday	10:00 PM

Contacts *can* be made on non-activity nights as well, especially if the band is open. It may just take longer to get someone's attention.

Local VHF/UHF nets often meet during activity nights. Two national organizations, SMIRK (Six Meter International Radio Klub) and SWOT (Sidewinders on Two), run nets in many parts of the country. These nets provide a meeting place for active users of the 50- and 144-MHz bands. For information on the meeting times and frequencies of the nets run by SMIRK or SWOT, ask other occupants of the bands in your area or see the Resources and References Guide for more information.

What's That Beeping?

If you tune around on 6 meters when propagation conditions are good, you'll probably hear several beacon stations. Beacons send their call signs and other information in slow-speed Morse code. Other information may include their grid-square locations (see below), output power and antenna height. Most beacons use non-directional antennas and relatively low power. If you can hear a beacon in a certain geographic area, you can probably work stations in that area. If you hear a beacon signal from several hundred or several thousand miles away, the band is open. Point your beam toward the beacon and start calling CQ! (Don't call CQ on the beacon frequency, though.) There are beacons on the bands above 50 MHz also, and they operate in a similar fashion. An extensive list of VHF beacons appears in *The ARRL Operating Manual*.

Grid Squares

One of the first things you'll notice when you tune the low end of any VHF band is that most QSOs include an exchange of *grid squares*. Grid squares are a shorthand means of describing your general location anywhere on earth. (For example, instead of trying to tell distant stations that, "I'm in Canton, New York," I tell them, "I'm in grid square FN24kp." It sounds strange, but FN24kp is a lot easier to locate on a map than a small town.)

Grid squares are coded with a 2-letter/2-number/2-letter code (such as FN24kp). This handy designator uniquely identifies the grid square and your exact location in latitude and longitude; no two have the same identifying code. There are several ways to find out your own grid square identifier. The ARRL offers a grid-square map of North America, a World Grid Locator Atlas and a program for PC-compatible computers (*GRIDLOC*). See the Resources and References Guide for more information.

Propagation

If you're new to the world above 50 MHz, you might wonder what sort of range is considered "normal." To a large extent, your range on VHF is determined by your location and the quality of

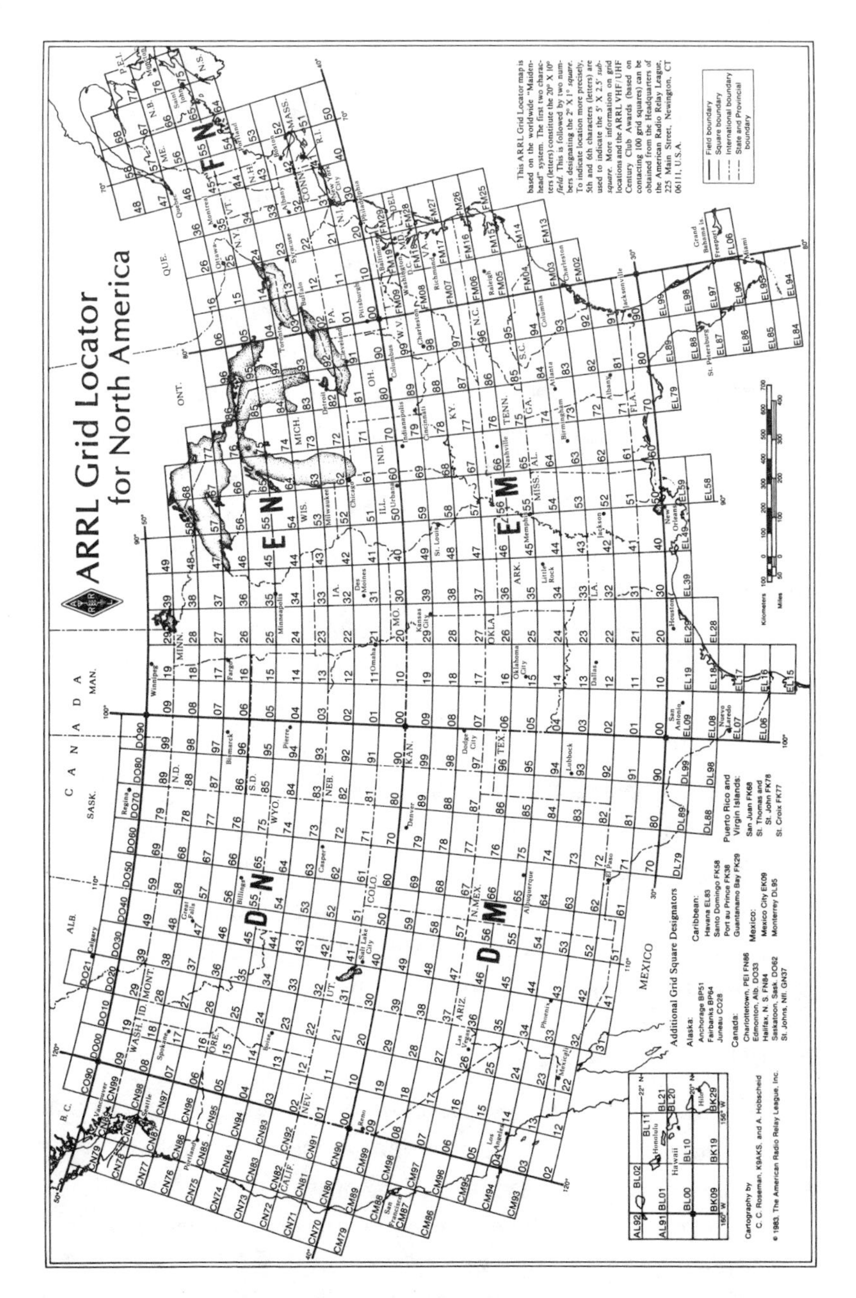

Fig 4-3—ARRL Grid Locator Map. You can use a map like this to determine your present grid location as well as the grid locations of other stations.

your station. Except during major propagation enhancements, a kilowatt and stacked beams at 100 feet will out-perform a 10-watt rig and a small antenna on the roof.

For the sake of discussion, consider a more-or-less "typical" station. On 2-meter SSB, a typical rig would be a low-powered, multimode transceiver (SSB/CW/FM), followed by a 100-watt amplifier. The antenna of our typical station might be a single 15-element Yagi at around 50 feet, fed with low-loss coax.

Using SSB or CW, how much territory could this station cover on an average evening? Location plays a big role, but it's probably safe to say you could talk to similarly equipped stations about 200 miles away almost 100% of the time. Naturally, higher-powered stations with high antennas and low-noise receivers have a greater range, up to a practical maximum of about 350-400 miles in the Midwest (less in the hilly West and East).

On 222 MHz, a similar station might expect to cover about the same distance, and somewhat less (perhaps 150 miles) on 432 MHz. This assumes normal propagation conditions and a reasonably unobstructed horizon. This range is a lot greater than you would get for noise-free communication on FM, and it represents the sort of capability the typical station should seek. Increase the height of the antenna to 80 feet and the range might extend to 250 miles, and probably more, depending on your location. That's not bad for reliable communication!

Band Openings and DX

The main thrill of the VHF and UHF bands for most of us is the occasional band opening, when signals from far away are received as if they are next door! DX of well over 1000 miles on 6 meters is commonplace during the summer, and occurs at least a few times each year on 144, 222 and 432 MHz.

DX propagation on the VHF/UHF bands is strongly influenced by the seasons. Summer and fall are definitely the most active times in the spectrum above 50 MHz, although band openings occur at other times as well. Here is a review of the most popular types of VHF/UHF propagation. Remember that there is a lot of variation, and that no two band openings are alike. This uncertainty is part of what makes VHF/UHF interesting and fun!

- *Tropospheric—or simply "tropo"—openings.* Tropo is the most common form of DX-producing propagation on the bands above 144 MHz. It comes in several forms, depending on local and regional weather patterns. This is because it is caused by the weather. Tropo may cover only a few hundred miles, or it may include huge areas of the country at once. The best times of year for tropo propagation are from spring to fall, although they can occur anytime.

  One indicator of a possible tropo opening is dew on the grass in the evening. Another is a high-pressure weather system stalled over or near your location.
- *Meteor scatter* communication uses the ionized trails meteors leave as they pass through the ionosphere's E layer. This ionization lasts only a second or so. The ionization is so intense, though, that even 432-MHz signals can sometimes be refracted. Most meteor-scatter QSOs are made on 6 and 2 meters. Because the ionization from a single meteor is brief, special operating techniques are used.

  Meteor-scatter contacts are possible at any time of year. Activity is greatest during the major meteor showers, especially the Perseids, which occurs in August.
- *Sporadic E* (abbreviated E_S) propagation is the most spectacular DX producer on the 50-MHz band, where it may occur almost every day during late June, July and early August. A short E_S season also occurs during December and January. Sporadic E is more common in mid-morning and again around sunset during the summer months, but it can occur at any time and any date. E_S also occurs at least once or twice a year on 2 meters in

most areas. E_S results from small patches of ionization in the ionosphere's E layer. E_S signals are usually strong, but they may fade away without warning.

- *Aurora* (abbreviated Au) openings occur when the auroras are sufficiently ionized to refract radio signals. Auroras are caused by the earth intercepting a massive number of charged particles from the Sun during a solar storm. Earth's magnetic field funnels these particles into the polar regions. The charged particles often interact with the upper atmosphere enough to make the air glow. Then we can see a visual aurora. The particles also provide an irregular, moving curtain of ionization that can propagate signals for many hundreds of miles.

  Aurora-reflected signals have an unmistakable ghostly sound. CW signals sound hissy, SSB signals sound like a harsh whisper. That's because the rapidly moving auroral curtain is modulating the signals as it refracts them. FM signals refracted by an aurora are unreadable.
- *EME, or Earth-Moon-Earth* (often called moonbounce) is the ultimate VHF/UHF DX medium. Moonbouncers use the moon as a passive reflector for their signals, and QSO distance is limited only by the diameter of the earth (both stations must have line of sight to the moon). Moonbounce QSOs between the USA and Europe or Japan are commonplace—at frequencies from 50 to 10,368 MHz. That's DX!

  Hundreds of EME-capable stations are now active, some with gigantic antenna arrays. Their antenna systems make it possible for stations running 100 watts and one or two Yagis to work them. Activity is constantly increasing. The ARRL sponsors an EME contest, in which moonbouncers compete on an international scale.

Hilltopping and Portable Operation

Maybe you'd like to try VHF/UHF weak-signal operating,

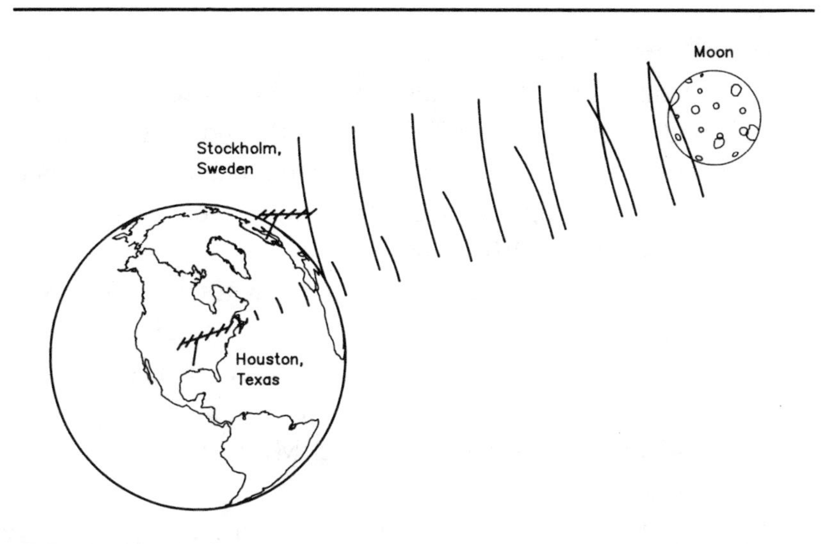

Fig 4-4—By reflecting its signal off the surface of the moon, a station in Houston, Texas is able to contact a station in Stockholm, Sweden. This is known as moonbounce, or EME. In addition to the equipment requirements discussed in this chapter, *both* stations must have a clear line-of-sight path to the moon.

but can't put up a tower. Or, you may live in a valley, from which propagation is poor. The solution? Take to the hills! VHF/UHF antennas are relatively small, and station equipment can be packed up and easily transported. Portable operation, commonly called hilltopping or mountaintopping, is a favorite activity for many amateurs. During VHF and UHF contests, a station located in a rare grid square is very popular. If you're on a hilltop or mountaintop, you'll have a very competitive signal.

Start by setting up on an easily accessible hill or mountain for an afternoon during a contest period. For a first effort, just take along a 2-meter rig. Even if you have an FM-only rig, you can still participate. Use the common simplex frequencies, like 146.52 and 146.55 MHz. If you find that the location "plays," you know where to take your new multimode rig next time!

VHF Contesting

Amateur Radio contests test your ability to work the most stations in different geographical areas on the most bands during the contest period. Contests also give you a chance to evaluate your equipment and antennas, and to compare your results with others. In most VHF/UHF contests, each contact is worth a certain number of points. You multiply your point total by the total number of different grid squares (*multipliers*) to calculate your final score. The only restrictions in these contests are that contacts through repeaters (and satellites) don't count, and the national 2-meter FM calling frequency, 146.52 MHz, is off limits for ARRL contest QSOs.

During the first hour or two of a VHF contest, contacts may come fast and furious. At other times, VHF contesting is more like an extended activity hour. VHF contests provide set times during which many other stations are operating. The concentrated activity gives you a chance for many contacts. During a contest, you'll know right away if there's a band opening!

Depending on your location, you may be able to work dozens of different grid squares on several bands, which makes for a high score and lots of fun. If you're interested in awards chasing, you'll also be pleased to know that many hams travel to rare grid squares for contests.

Who Can Enter?

Most VHF/UHF contests are open to any licensed amateur who wants to participate. The ARRL sponsors all the major VHF/UHF contests (see Table 4-2), and specific rules, descriptions of the different categories, as well as entry forms, are available from ARRL Headquarters. You don't have to be an ARRL member to participate in these contests, nor are you required to submit your logs, although doing so helps the ARRL Contest Branch verify QSOs that others claim.

VHF/UHF contests feature a variety of categories among which you can choose. For single operators (those operating without assistance), entry classes in the ARRL contests include all-band, single-band, QRP portable, and one for Rovers (those who operate from more than one grid square during the contest). *The ARRL Operating Manual* is a good source of more information on selecting an entry category, and what it takes to go QRP portable or roving to different grid squares during a contest.

When and Why?

The ARRL VHF contests are held throughout the year, with emphasis on the warmer months to encourage hilltop operation during favorable weather (Fig 4-5). Outside of that, the ARRL VHF/UHF contest program is designed to take the best advantage

Table 4-2

Major ARRL VHF Contests

Contest	*Bands*	*Typical Propagation Enhancements*
January VHF Sweepstakes	50 MHz and up	Aurora, meteor scatter
Spring Sprints	One per band	Varies
June VHF QSO Party	50 MHz and up	Sporadic E
August UHF Contest	222 MHz and up	Tropospheric ducting
September VHF QSO Party	50 MHz and up	Tropospheric ducting, aurora, meteor scatter
EME Contest	All amateur bands*	

*EME contests are scheduled for weekends with the most favorable conditions for EME communications. (See *The ARRL Operating Manual* for details.)

Fig 4-5—As high as you can get in New England: ARRL Laboratory Engineer Zack Lau, KH6CP, operating from Mt Washington, New Hampshire.

of propagation enhancements that usually occur at certain times of the year. For instance, the June VHF QSO Party almost always occurs during periods of excellent sporadic-E propagation, giving you an opportunity to enjoy long-distance contacts on 6 and 2 meters. In fact, the first documented sporadic-E contact on the 222-MHz band was made during a June VHF QSO Party. Contests are more than recreational; they also give amateurs the chance to contribute to radio science.

As shown in Table 4-2, the major ARRL VHF contests consist of the January Sweepstakes, June and September VHF QSO Parties, August UHF Contest, VHF/UHF Spring Sprints, and the International EME (moonbounce) Contest. Except for the Sprints, these events encompass many bands each. The January SS and June and September QSO Parties are the most popular of them all, and each permit activity on SSB, CW and FM on all amateur frequencies from 50 MHz and up.

The UHF Contest is slightly different than the other contests described so far. The major difference is that only QSOs on the 222-MHz and higher bands are allowed for contest credit. The Spring Sprints are single-band, 4-hour contests held over a several-week-long period. Sprints occur on the appropriate activity night for each band, so most Sprints are held on week nights. These short contests provide a super opportunity to test a new location or piece of equipment.

When To Be Where

You'll find lots of random 6- and 2-meter activity during VHF contests. FM is rare on 6 meters in the US, even during contests, but it's quite common in most areas on 146, 222 and 440 MHz. On SSB, most stations stay near the calling frequencies of 50.125, 50.200, 144.200, 222.100 and 432.100. On CW, look between 80 and 100 kHz above 50, 144, 222 and 432 MHz. (Six meters offers less CW activity than the other VHF/UHF bands.) On FM, check the simplex calling frequencies (listed in *The ARRL Repeater Directory*), *except* 146.52 MHz.

Most contest stations try to be on the appropriate bands during the established activity hours. Because the major contests last 33 hours, you have three shots at each band (Saturday evening, Sunday morning and Sunday evening).

Transceivers

Aside from antennas, the one necessity of radio communication is a radio! This section introduces available equipment and your alternatives in equipping your VHF station. In general, transceivers are the easiest way to get on VHF. *Transverters*—self-contained receive and transmit converters designed to go with an HF rig—run second in ease of use, but often first in performance.

What's Available?

Multimode VHF transceivers can be grouped into two distinct classes: home station and mobile/portable. A look at *The ARRL Radio Buyer's Sourcebook* will help you decide what's right for you. It's also a good idea to review recent *QST* product reviews when you're selecting equipment.

Many people just getting into VHF settle on multimode, single-band mobile or portable transceivers (Figs 4-6 and 4-7). These rigs are often less expensive, less complex and more flexible (in terms of power sources and size) than home-station rigs. Some home-station rigs include accessories not usually found in portable and mobile rigs. Serious operators find these accessories, such as preamplifiers, narrow IF filtering

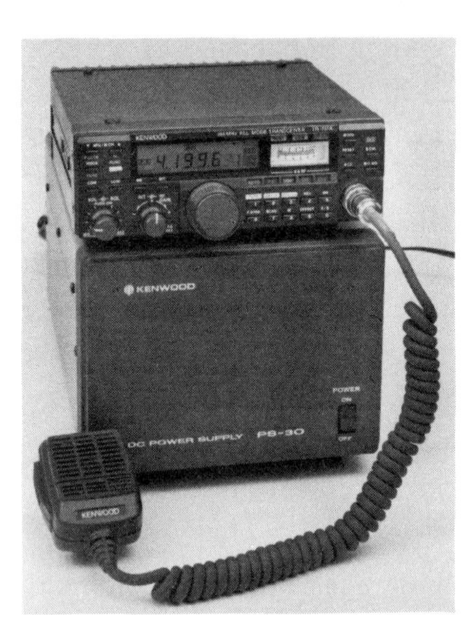

Fig 4-6—The Kenwood TR-751A is an "all-mode" 2-meter transceiver. In addition to FM (or packet radio), it allows operation on CW, SSB and AM. This 25-watt-output transceiver is no larger than most FM-only transceivers in its power class. The PS-30 13.6-volt power supply on which the transceiver is sitting is suitable for home-station use.

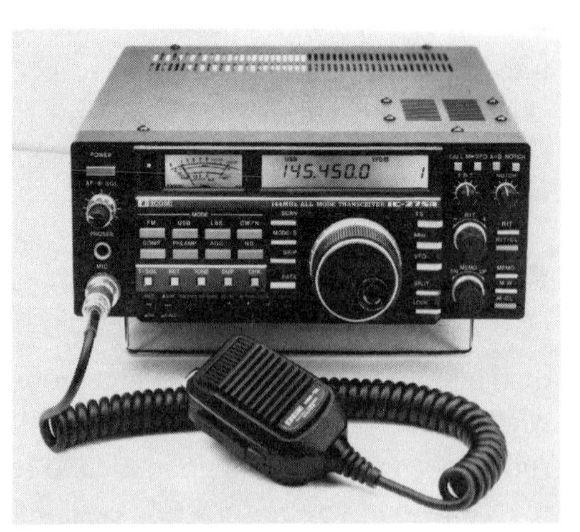

Fig 4-7—ICOM's IC-275A transceiver is similar to the Kenwood TR-751A shown in Fig 4-6. One difference is a built-in power supply for home-station use. The IC-275A can also be used with an external 13.6-V dc supply for mobile or portable operation.

and noise blankers helpful when propagation and interference conditions make it hard to hear another station.

Although most VHF multimode transceivers are single-band radios, multiband transceivers have been growing in popularity (Figs 4-8 and 4-9). Usually aimed at the amateur satellite market,

Fig 4-8—Yaesu's FT-736R multimode transceiver allows FM/packet, SSB, CW and AM operation from 144-148 MHz and 430-450 MHz. You can install two additional modules to cover any combination of 50-54 MHz, 222-225 MHz and 1240-1300 MHz. In addition to a built-in ac power supply, the FT-736R includes many features (such as IF shift, notch filter and speech processor) normally found only on HF transceivers. Features attractive to amateur-satellite users are also standard.

Fig 4-9—The Kenwood TS-790A multimode VHF/UHF transceiver. Coverage of the 144- and 430-MHz bands is standard; a 1240- to 1300-MHz module is optional. The TS-790A also includes many features normally found only on HF transceivers, plus the ability to receive on two bands at once.

these rigs are also popular among terrestrial operators because of their flexibility. They usually allow you to receive on one band while transmitting on another. These rigs are considerably more expensive than their single-band counterparts, but less expensive than buying separate radios for each band they cover.

Transverters

An alternative to buying one or more VHF transceivers is to buy or build a transverter to go along with your HF rig. Although this equipment sometimes requires some effort to interface with an HF rig (except for those made to go with your particular transceiver), the performance and cost savings can be substantial.

With modern components, you might be surprised at how few parts and how little cost and expertise it takes to get your own gear on the air. A *QST* article by Ed Krome, KA9LNV ("A High-

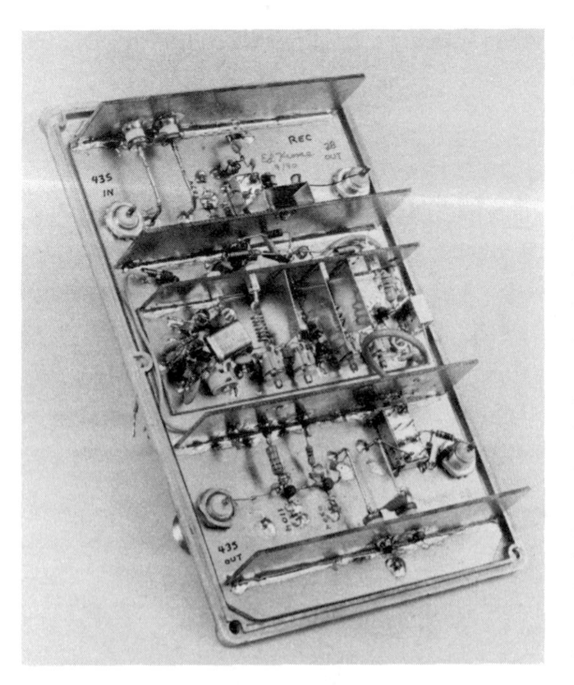

Fig 4-10—Inside view of a *transverter* built by Ed Krome, KA9LNV. It converts the transmitter output of a 10-meter transceiver to the 430-MHz band; 430-MHz signals are converted to 10 meters. Complete construction details for the transverter and a companion power amplifier appeared in *QST* (see text). If you know how to solder, you can build this transverter easily.

Performance, Easy-to-Build 432-MHz Transverter," *QST*, August and September 1991), is a good example. Ed's 432 transverter is shown in Fig 4-10.

Steer clear of anything made before 1980 or so, because equipment performance and reliability have come a great distance since then. Older gear may not meet your needs for very long.

Antennas

When you're assembling a VHF station, one of your first priorities is selecting antennas. Should you use verticals or beams? Should you buy or build your antennas? Before answering these questions, let's have a look at your choices.

What's Out There?

VHF antennas can be grouped into three broad categories: verticals, multielement directive antennas (such as Yagis and quads), and reflectors. Figs 4-11, 4-12 and 4-13 illustrate the three types. By far, the most commonly used VHF antennas fall into the first two categories.

Vertical antennas are popular for FM and packet operation. They provide *omnidirectional* coverage (radiate equally in all directions), and are useful for most casual operation. Almost all weak-signal VHF/UHF work, though,

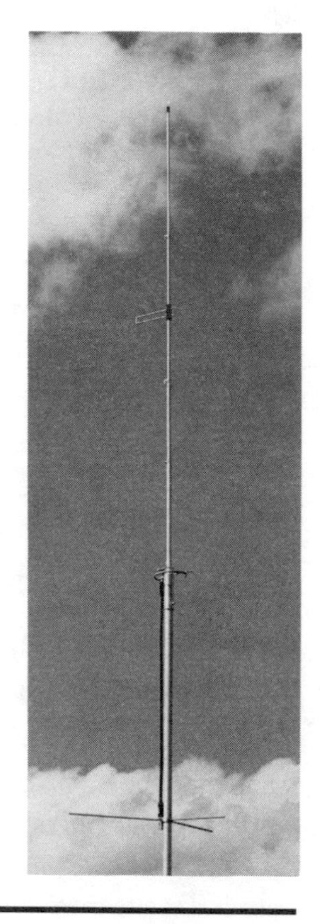

Fig 4-11—Cushcraft's ARX-2B Ringo Ranger vertical is a popular FM and packet antenna.

Fig 4-12—AA4FQ used this array in an ARRL June VHF Contest. From top to bottom: 13-element 2-meter beam; pair of 24-element 432-MHz beams; 7-element 6-meter beam. ***(AA4FQ photo)***

Fig 4-13—WA8NLC uses this 17-foot parabolic ("dish") antenna for 1296-MHz EME operation.

is done with horizontally polarized, directional antennas.

The Yagi is the most common directive antenna. Yagi antennas are commercially available with three to at least 33 elements. Examples of some popular commercial Yagis are shown in Figs 4-14 and 4-15.

Two other kinds of multielement directional antennas are shown in Figs 4-16 and 4-17. One is a quad, which uses loop elements instead of wire or rod elements, and the other is a log-periodic dipole array (LPDA), usually referred to simply as a log periodic. In terms of performance on a single band, quads are basically the same as Yagis. LPDAs, on the other hand, cover very wide frequency ranges. The antenna shown in Fig 4-17 covers 50 to 1300 MHz—that's six ham bands and everything in between! The penalty for this frequency coverage is significantly lower gain than can be achieved with a single-band Yagi or quad. They're also considerably more complex than Yagis and quads.

Fig 4-14—The Cushcraft 17B2 2-meter beam is a *big* antenna! Two struts reinforce the boom, allowing it to be made of lighter-gauge aluminum.

Fig 4-15—An array of 432-MHz Yagi antennas designed and built by Steve Powlishen, K1FO. Steve has installed these antennas so they can be rotated and tilted upward, for EME communications. Although these 22-element beams are fairly long (about 14 feet), the booms are built in three pieces. It's easy to take one or two on a portable contest expedition, and assemble them on the spot. At home, they'll store easily in your basement or garage.

Fig 4-16—A portable 2-meter quad antenna. The elements fold back on each other for transport. Construction of this antenna is described in *The ARRL Antenna Book*.

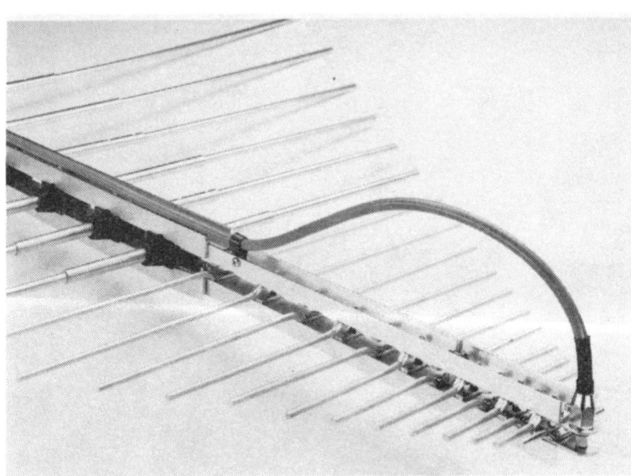

Fig 4-17—Part of the Creative Design CLP5130-1 VHF/UHF log-periodic antenna. With a boom length of less than 6 feet, this antenna covers *all* frequencies from 50 to 1300 MHz! Although this antenna's performance is not equal to a stack of single-band beams, it's ideal for hams who can only install one antenna.

On the other hand, the convenience of having coverage of so many bands with only one antenna and feed line is very attractive, especially for portable operation.

One other antenna that deserves mention is the quagi. A hybrid of Yagi and quad designs, the quagi uses quad-loop elements for the driven element and reflector and Yagi-type directors. Quagis are somewhat easier to build from scratch than long Yagis, yet offer good performance.

How Do I Choose?

There's even more variety in antennas than in equipment for the VHF bands. Unless your local dealer carries all the major brands, you'll do most of your shopping without actually seeing all available antennas up close. That's why it's a good idea to get catalogs from major retailers who advertise in *QST*, which show at least the major specifications for the antennas they carry. Also, read *QST* Product Reviews, and ask your friends what they're using.

Before you can select an antenna system, decide what you want to do on VHF (FM packet and/or repeaters, ATV, weak-signal CW and SSB and so on). For FM voice and packet, vertical polarization is the norm, so a vertical or short Yagi (or one for each band) may serve you well. SSB, CW, ATV and other modes where you're more interested in DXing or communicating under challenging conditions, require high-gain, directive antennas. Yagis are the most common (and easiest to work with) of these.

Very few antennas go right from box to mast; for the most part, you'll be building every antenna you use. If you're considering home-made antennas, you can build verticals and small, simple Yagis and quagis. More elaborate antennas require a good collection of tools, test equipment and some metalworking experience.

Antenna Supports

Most VHF antennas are small and light enough to be supported and rotated with standard TV-antenna hardware. You don't need a huge tower and mast for casual VHF work. For successful long-distance communication, you'll want to consider a tall support and high-gain, highly directive antennas.

Vertical antennas require even less support hardware than multielement antennas. A look through the Radio Shack catalog section on TV masts, brackets and hardware will show you many of your options.

Feed Lines: The Weak Link

When you install your antennas, you'll need to connect them to your radios via feed lines. No surprise so far, right? What makes this subject worth discussing here is *loss*. Cable loss is a function of conductor losses. Physically large cables have lower loss than smaller cables, because they have more conductor area. That's why you'll hear people talking about feeding VHF antennas with *Hardline* and *Heliax* (types of rigid and semirigid coax).

How does this affect you? Well, in practical terms, you shouldn't use small cables like RG-58 and RG-8X *at all* at VHF. RG-8 and RG-213 are acceptable for short runs (under 50 feet or so at 432, under 100 feet at 6 meters). For longer runs, consider more expensive cables like Belden 9913 or an equivalent—they don't cost much more than RG-213, and the precious decibels you'll save are well worth the extra cost.

There's no sense spending your money on quality equipment and antennas, and then wasting it all in lossy coax. It follows that you should keep all cable runs as short as possible. If you must run very long cables to your VHF antennas, consider moving up to higher-grade cables. You can find more details in *The ARRL Handbook* and *The ARRL Antenna Book*.

The Challenge and the Reward

There's no question that it's easier to get on the air with FM than SSB or CW. With FM, it may be a matter of simply buying a hand-held transceiver and talking through your local repeater.

SSB and CW take a little more effort, but the reward is considerable!

As a weak-signal operator, you'll enjoy contacts over distances that FM enthusiasts can only achieve through complex linked-repeater systems. Best of all, you'll experience the true magic of VHF operating. As you sharpen your skills, you'll be able to predict when band openings are about to take place. By listening to the distant signals, you'll know which propagation mode is active and how to use it to your advantage.

Weak-signal VHF operating will challenge you every day. DX stations sometimes appear when you least expect them—and disappear just as suddenly. Wait until the day when you turn on your equipment and hear a flood of distant CW and SSB signals. The excitement will be electrifying and you'll know in that moment what you've guessed all along: there is much more to VHF than FM!

CHAPTER 5

Satellite Communications

By Steve Ford, WB8IMY

It's incredible to think that amateur satellites have been in orbit since the early '60s. Even before astronaut John Glenn made his historic flight, OSCAR 1 (*O*rbiting *S*atellite *C*arrying *A*mateur *R*adio) was circling the earth, transmitting "HI" in CW.

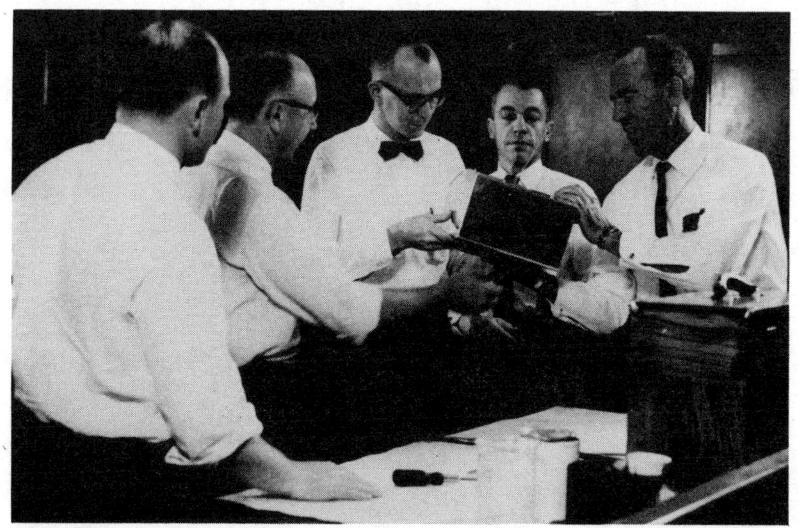

OSCAR 1, the first Amateur Radio satellite, was designed and built by a small group of ham radio operators in California. It was launched as ballast by the Air Force in December 1961.

Today you can choose from a variety of extremely sophisticated amateur satellites. You can even communicate with Russian cosmonauts and American space shuttle astronauts. What may surprise you more than anything else, however, is the *ease* with which you can access most of these satellites. It seems to be one of the best kept secrets in Amateur Radio!

On the Wings of a DOVE

Do you own a 2-meter FM rig? How about an outdoor antenna of some kind? Excellent! You are now the proud owner of a basic satellite receiving system—and there is a bird in orbit just waiting to talk to you. It's called DO-17, otherwise known as *DOVE.*

DOVE Telemetry

Raw data packets and decoded information

```
                    DOVE  DAYTIME  TELEMETRY
  RAW DATA
DOVE-1>TLM:00:5A 01:5A 02:88 03:32 04:59 05:58 06:6C 07:4A 08:6C 09:68 0A:A2
0B:EC 0C:E8 0D:DC 0E:3F 0F:24 10:D8 11:93 12:00 13:D1 14:9B 15:AE
16:83 17:7C 18:76 19:7E 1A:7C 1B:45 1C:84 1D:7B 1E:C4 1F:6C 20:CF
DOVE-1>TLM:21:BB 22:79 23:26 24:22 25:26 26:01 27:04 28:02 29:3A 2A:02 2B:73
2C:01 2D:7C 2E:58 2F:A2 30:D0 31:A2 32:17 33:6B 34:AC 35:A2 36:A6
37:A8 38:86 39:A2 3A:01
DOVE-1>STATUS: 80 00 00 85 B0 18 77 02 00 B0 00 00 B0 00 00 00 00 00 00 00
DOVE-1>BCRXMT:vary= 21.375 vmax= 21.774 temp=  7.871
DOVE-1>BCRXMT:vbat= 11.539 vlo1= 10.627 vlo2= 10.127 vmax= 11.627 temp=  3.030
DOVE-1>WASH:wash addr:26c0:0000, edac=0x61
DOVE-1>TIME-1:PHT: uptime is 086/01:14:32.  Time is Sat Mar 10 15:43:26 1990

  DECODED TELEMETRY

DOVE   uptime is 086/01:14:32.  Time is Sat Mar 10 15:43:26 199

Rx E/F Audio(W)     2.21 V  Rx E/F Audio(N)     2.21 V  Mixer Bias V:       1.39 V
Osc. Bisd V:        0.51 V  Rx A Audio (W):     2.19 V  Rx A Audio (N):     2.16 V
Rx A DISC:          0.41 k  Rx A S meter:      74.00 C  Rx E/F DISC:       -1.08 k
Rx E/F S meter:   104.00 C  +5 Volt Bus:        4.94 V  +5V Rx Current:     0.02 A
+2.5V  VREF:        2.51 V  8.5V BUS:           8.60 V  IR Detector:       63.00 C
LO Monitor I:       0.00 A  +10V Bus:          10.96 V  GASFET Bias I:      0.00 A
Ground REF:         0.00 V  +Z Array V:        21.38 V  Rx Temp:            7.26 D
+X (RX) temp:      -4.24 D  Bat 1 V:            1.35 V  Bat 2 V:            1.36 V
Bat 3 V:            1.38 V  Bat 4 V:            1.34 V  Bat 5 V:            1.37 V
Bat 6 V:            1.57 V  Bat 7 V:            1.36 V  Bat 8 V:            1.37 V
Array V:           21.32 V  +5V Bus:            5.30 V  +8.5V Bus:          8.85 V
+10V Bus:          11.54 V  BCR Set Point:    131.48 C  BCR Load Cur:       0.18 A
+8.5V Bus Cur:      0.06 A  +5V Bus Cur:        0.17 A  -X Array Cur:      -0.01 A
+X Array Cur:      -0.00 A  -Y Array Cur:      -0.01 A  +Y Array Cur:       0.12 A
-Z Array Cur:      -0.01 A  +Z Array Cur:       0.25 A  Ext Power Cur:     -0.02 A
BCR Input Cur:      0.45 A  BCR Output Cur:     0.29 A  Bat 1 Temp:         3.02 D
Bat 2 Temp:       -24.81 D  Baseplt Temp:       3.02 D  FM TX#1 RF OUT:     0.05 W
FM TX#2 RF OUT:     0.97 W  PSK TX HPA Temp    -3.03 D  +Y Array Temp:      3.02 D
RC PSK HPA Temp     0.60 D  RC PSK BP Temp:    -0.61 D  +Z Array Temp:     19.97 D
S band HPA Temp     3.02 D  S band TX Out:     -0.04 W
```

Fig 5-1—Typical telemetry received and decoded from the DOVE satellite.

DOVE is one of several *Microsats* presently in orbit. They're called Microsats because of their tiny size (9 inches on each side). Its primary mission is education. DOVE transmits streams of packet telemetry and occasional bulletins on 145.825 MHz. By studying the telemetry, you can learn all sorts of fascinating things about conditions in space (see Fig 5-1). Since DOVE is a *LEO* (*L*ow *E*arth *O*rbiting) satellite, its signal is very easy to hear.

If you only want to listen, you'll get an earful of raucous packet bursts as it streaks overhead. DOVE also has digital voice capability and may be transmitting in that mode from time to time.

If you have packet equipment, you're in for an extra treat. Set up your TNC as you would for normal operation and switch your FM transceiver to 145.825 MHz. As DOVE rises above the horizon you'll begin to see streams of data flowing across your monitor. You may also see brief text bulletins.

After you get tired of watching raw data, you'll want to find out what it means. There are several programs available to decode DOVE telemetry. See the Resources and References Guide for more information.

DOVE has had a troubled history, with several failures during its career. Each time, however, AMSAT volunteers have managed to revive the satellite and get it back into working order. When DOVE is operating, it pumps out a strong signal. I've heard it clearly on a hand-held transceiver with just a rubber duck antenna.

The *Mir* Space Station

You've discovered how easy it can be to eavesdrop on satellite signals. Now it's time to start thinking in terms of *transmitting* to a satellite. A perfect candidate for your first QSO is the Russian *Mir* space station.

Mir has been occupied by Russian cosmonauts for several years as a laboratory for testing human responses to long-duration space flights. The *Mir* studies are extremely important for future

Finding the Satellites

Amateur Radio satellites are in non-geostationary orbits. This simply means that the satellites are not in fixed positions in the sky from our perspective here on Earth. They are like tiny moons, rising and setting rapidly over your local horizon. While the satellites zip around the Earth at tremendous speeds, the Earth is turning beneath them. The result is that you can't rely on satellites to appear in the same places at the same times each day.

Orbital Elements

So how can you know when a satellite is about to make an appearance in your neighborhood? To answer that question you need to know the satellite's *orbital elements.* Take a look at the elements shown in Table 5-1.

An orbital element set is merely a collection of numbers that describes the movement of an object in space. By feeding the numbers to a computer program, you can determine exactly where a satellite is (or will be) at any time. Don't worry about the definitions of *Mean anomaly, Argument of perigee* and so on. If you're curious, get a copy of the *Satellite Experimenter's Handbook* (see the Resources and References Guide for more information) and you'll learn all about those definitions—and more. For the moment, consider the words as labels for the numbers that appear beside them.

Finding the Elements

There are several sources for orbital elements:

- ❑ Satellite newsletters
- ❑ W1AW RTTY and AMTOR bulletins
- ❑ Packet bulletin boards
- ❑ Telephone bulletin boards
- ❑ AMSAT nets

(See the Resources and References Guide for more information.)

If you have an HF radio, RTTY capability, a packet TNC, a telephone modem or the necessary cash for a subscription, you'll always be able to get the latest orbital elements for the satellites you want to track. If all else fails, there is probably someone in your area who has access to the elements. Ask around at your next club meeting.

Using the Elements

Computers are common in most Amateur Radio stations

Table 5-1

Example of Orbital Elements Provided for Computer Tracking Programs

Parameter Name	*Value*	*Units*
Satellite:[1]	OSCAR 13	
Catalog Number:	19216	
Epoch Year:	1988	
Epoch Time:	258.28144	days
Element Set:	RUH9-88	
Inclination:	57.57	deg
RAAN:[2]	239.56	deg
Eccentricity:	0.6563	
Arg of Perigee:[3]	190.53	deg
Mean Anomaly:	0.0	deg
Mean Motion:	2.09699369	rev/day
Decay Rate:[4]	0.	rev/day/day
Epoch Rev:[5]	193	
Semi-Major axis:	25783	km
Bahn Latitude[6]	0	deg
Bahn Longitude[7]	180	deg

Alternate parameter names

[1]Object
[2]Right Ascension of Ascending Node; RA of Node
[3]Argument of Perigee
[4]Drag factor; Rate of change on mean motion, first derivative of mean motion
[5]Revolution number, orbit number
[6]ALAT, BLAT
[7]ALON, BLON

today. If you have a computer in your shack, you're in luck! There are many programs on the market that will take your orbital elements and magically produce satellite schedules.

Among other things, the programs tell you when satellites will appear above your local horizon and how high they will rise in the sky (their elevation). When working satellites, the higher the elevation the better. Higher elevation means less distance between you and the satellite with less signal loss from atmospheric absorption.

Some programs also display detailed maps showing the *ground track* (the satellite's path over the ground). AMSAT (the Radio Amateur Satellite Corporation) offers satellite tracking software for a variety of computers. For more information contact: AMSAT, PO Box 27, Washington, DC 20044, tel 301-589-6062.

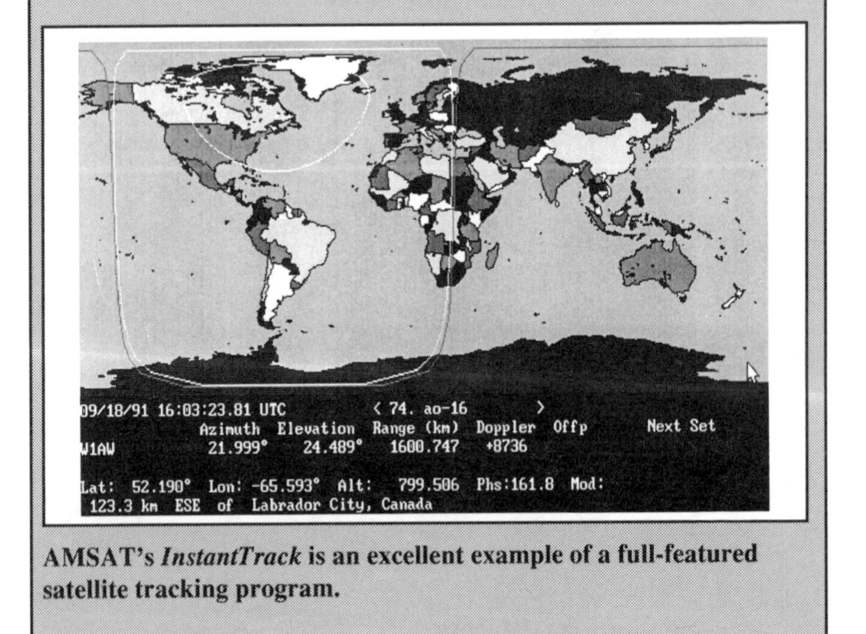

AMSAT's *InstantTrack* is an excellent example of a full-featured satellite tracking program.

manned missions to Mars and beyond.

To combat boredom, an Amateur Radio station was installed. The cosmonauts pass amateur license tests and are assigned special *Mir* call signs (such as U8MIR) prior to launch.

When they reach the station, they operate 2-meter FM voice or packet.

Like the DOVE satellite, *Mir*'s signal is powerful. You'll usually find it on 145.55 MHz, and you won't need sophisticated equipment to hear it—or to be heard. Once again, an outside antenna—such as a ground plane—works fine. Its orbit provides a couple of very good "passes" each day for most areas.

Mir on Packet

The *Mir* Amateur Radio station uses standard 1200-baud AFSK packet—the same packet format you use for QSOs here on earth. The *Mir* packet station includes a mailbox where you can leave messages for the cosmonauts (or anyone else) and pick up their replies.

The biggest problem with working *Mir* on packet is interference—lots of interference! With the signal coverage the space station enjoys, you can imagine how many hams might be trying to connect to *Mir* at the same time. This creates pure chaos as far as its FM receiver is concerned.

If you're able to connect to the mailbox, the constant bombardment of signals may make it difficult for you to post your message. (Remember that you may only have a few minutes before the space station slips below your horizon.) Here are a couple of tips to improve your chances:

- ❑ Listen before you start sending your connect requests. Monitor a few transmissions and make sure you have the correct call sign. The call sign changes whenever a new crew occupies the station.
- ❑ Use as much power as you have available. If there were only a couple of stations competing for *Mir*'s receiver, you'd only need a couple of watts to have a decent chance of connecting. During a normal pass, however, there are usually *dozens* of stations blasting out connect requests. The stations that pack the

bigger punches seem to win consistently.

- ❑ Try connecting during "unpopular" hours. If you have the stamina to sit up and wait for a late-night pass, you may have a better opportunity to make a connection.

When you finally connect to the mailbox, make your message entry *short*. The station will be out of range before you know it and other hams will be waiting to try their luck. Some packet software permits the user to create a message file before attempting to connect. If your software offers this feature, it will come in handy for *Mir*.

Voice Contacts with Mir

The *Mir* cosmonauts obviously enjoy packet, but sometimes they crave the sounds of other human voices. You may be waiting for a chance to connect on packet, only to hear them calling CQ instead!

Working *Mir* on voice is very similar to working a DX pile-up. You sit with microphone in hand and wait until you hear the cosmonaut complete an exchange. At that moment you key the mike and say your call sign. Now listen. No response? Call again quickly! Keep trying until you hear him calling you or someone else.

I've heard of hams working *Mir* while mobile and some claim to have worked *Mir* with hand-helds. As you might imagine, *Mir* QSL cards are highly prized!

The major problem with working *Mir*—on voice or packet—is its erratic schedule. The cosmonauts have many daily assignments and are not always able to find the time to operate their amateur station. They are sometimes forced to turn off their equipment altogether to avoid interference to other systems during critical tests.

Another problem concerns *Mir*'s orbit. The space station travels at a relatively low altitude, so it's always subject to a

An actual QSL card verifying contact with the Russian *Mir* space station. *Mir* QSLs are processed and mailed *after* a cosmonaut has returned to earth. You may have to wait a while to receive a card like this one, but it's well worth it!

significant amount of atmospheric drag. If it didn't occasionally "boost" to a higher orbit, the station would reenter the atmosphere and be destroyed. Every time *Mir* fires its rocket engines to adjust its orbit, a revised set of orbital elements must be distributed. If you want to try your luck with *Mir*, plan to update your elements for the space station as often as possible.

As this book was being written, the former Soviet Union was undergoing severe political and economic turmoil. What effect this will have on the future of the *Mir* space station has yet to be determined. It is assumed that the station will continue to be occupied—and active on Amateur Radio. Even so, it's a good idea to check your Amateur Radio news sources (such as *QST*) for updates on the status of *Mir*.

SAREX

SAREX, the *S*huttle *A*mateur *R*adio *Ex*periment, is a continuing series of Amateur Radio operations from US space shuttle missions. The first *SAREX* operations employed 2-meter FM voice, but more recent flights have also used packet, ATV and SSTV. The variety of modes in use depends on the available cargo

A slow-scan television image of astronaut Tony England, WØORE, aboard space shuttle *Challenger* in 1985.

space. In addition, not every shuttle astronaut is a licensed ham, so not every shuttle mission has an active *SAREX* operation. Check *QST* for the latest news on upcoming *SAREX* missions.

Unlike *Mir*, *SAREX* uses a 600-kHz split-frequency scheme to accommodate standard 2-meter FM transceivers. Earthbound DXpeditions use split-frequency operation to maximize the number of stations they can work. The same is true for *SAREX*. (If operating from space isn't a DXpedition, I don't know what is!) With this thought in mind, you can appreciate the importance of knowing which frequencies are being used for the uplinks and downlinks. (They will be published in *QST* well in advance of the launch date.) Whatever you do, *never* transmit on the *SAREX* downlink frequency. This mistake will make you very unpopular very quickly!

You'll need the shuttle's orbital elements to predict when it will be in range and these will be available through the sources we've already discussed. You can use the same 2-meter equipment for *SAREX* as you do for *Mir*. During one of the "packet robot" operations on a previous flight, I managed to connect using my trusty ground plane. The shuttle was only 12° above my horizon at the time, but my signal still made it!

The RS (*Radio Sputnik*) Satellites

Have your ever heard of RS-10/11 or RS-12/13? They are unmanned Amateur Radio satellites placed in orbit by the former Soviet Union. Without a doubt they are among the easiest satellites to work.

The RS satellites are completely different from *DOVE*, *Mir* or the space shuttle. They are basically orbiting repeaters riding piggyback on larger satellite *platforms*. RS-10 and 11 are carried by *COSMOS* 1861, and RS-12 and -13 are part of *COSMOS* 2123. There are two RS satellites per platform (which is the reason for the dual designation), but only one satellite is active at a time. All RS satellites are equipped with unique devices called *linear transponders*.

Linear Transponders

Earthbound repeaters listen on one frequency and repeat what they hear on another. Imagine what would happen if your local repeater could retransmit everything it heard on an entire *group* of frequencies? This is exactly the function of a linear transponder.

In *mode A*, the RS satellite transponder listens to a portion of the 2-meter band and retransmits everything it hears on the 10-meter band. When *mode K* is active, the transponder listens to a section of the 15-meter band and simultaneously retransmits on 10 meters. In *mode T*, the satellite listens on 15 meters and retransmits on 2 meters! The range between the highest and lowest uplink (or downlink) frequencies is known as the transponder's *passband*. See the RS uplink and downlink passbands in Table 5-2.

Table 5-2

Linear Transponder Frequencies

RS Satellites

	RS-10	*RS-11*	*RS-12*	*RS-13*
Mode A				
Uplink	145.860-145.900	145.910-145.950	145.910-145.950	145.960-146.00
Downlink	29.360-29.400	29.410-29.450	29.410-29.450	29.460-29.500
Beacons	29.357/29.403	29.407/29.453	29.408/29.454	29.458/29.504
Mode A Robot				
Uplink	145.820	145.830	145.830	145.840
Downlink	29.357/29.403	29.407/29.453	29.454	29.504
Mode K				
Uplink	21.160-21.200	21.210-21.250	21.210-21.250	21.260-21.300
Downlink	29.360-29.400	29.410-29.450	29.410-29.450	29.460-29.500
Beacons	29.357/29.403	29.403/29.453	29.408/29.454	29.458/29.504
Mode K Robot				
Uplink	21.120	21.130	21.129 21.138	
Downlink	29.357/29.403	29.403/29.453	29.454 29.504	
Mode T				
Uplink	21.160-21.200	21.210-21.250	21.210-21.250	21.260-21.300
Downlink	145.860-145.900	145.910-145.950	145.910-145.950	145.960-146.000
Beacons	145.857/145.903	145.907/145.953	145.912/145.958	145.862/145.908
Mode T Robot				
Uplink	21.120	21.130	21.129 21.138	
Downlink	145.857/145.903	145.907/145.953	145.958 145.908	

Phase 3 Satellites

Satellite	*Mode*	*Uplink (MHz)*	*Downlink (MHz)*
AO-10	B	435.030-435.155	145.825-145.975
	Beacon		145.810
AO-13	B	435.420-435.570	145.825-145.975
	J	144.425-144.475	435.990-435.940
	L	1269.351-1269.641	435.715-436.005
	S	435.601-435.637	2400.711-2400.747
	Beacons		145.812, 145.985
			435.651
			2400.325

Other Satellites

Satellite	*Mode*	*Uplink (MHz)*	*Downlink (MHz)*
FO-20	J(A)	145.900-146.000	435.900-435.800
	Beacon		435.795
AO-21	B(1)	435.022-435.102	145.852-145.932
	B(2)	435.043-435.123	145.866-145.946
	Beacons(1)		145.822, 145.952

Not only do linear transponders repeat everything they hear on their uplink passbands, they do so very faithfully. CW is retransmitted as CW; SSB as SSB. FM voice transmissions are strongly discouraged since their broad signals occupy an enormous chunk of the downlink passband. Not only would this limit the number of stations that could use the satellite, it would place a severe drain on transponder power.

Working the RS Satellites

The first thing to do is to feed the RS orbital elements to your computer and find out when the best passes will occur in your area. As soon as you decide when you'll make your attempt, you have to determine which mode (and satellite) is active. Listening to the satellite itself will provide clues. You'll find that modes A and K are the most popular. On RS-10, mode A is usually available on a daily basis while mode K tends to be active Tuesday through Friday. (As this was being written, RS-12 had been locked into mode K on a continuous basis.) Mode T operation, on the other hand, is somewhat rare. Modes are often active *simultaneously*—such as modes KA and KT.

For VHF enthusiasts, mode A is the most popular. If you have a 2-meter SSB/CW rig and an 10-meter receiver, you're all set!

"I can receive on 10 meters, but I only have a 2-meter FM rig. I'll never be able to use mode A," the skeptic grumbles. On the contrary! Have you considered using your FM transceiver as a CW rig? All you have to do is wire a key to the push-to-talk (PTT) pins on a spare microphone plug. Your signal may be a little raw and "chirpy," but you'll be transmitting usable CW!

Elaborate antennas are definitely *not* required to work the RS satellites. A wire dipole is fine for receiving the 10-meter downlink signal. By the same token, a basic ground plane is adequate for your 2-meter uplink. In terms of power, 20 to 30

Finding Your RS Downlink Signal

If you know your uplink frequency, how can you predict where your signal will appear in the downlink passband? There is a calculation you can use that will get you in the ballpark:

$F_{Down} = F_T + F_{Up}$

F_T is the *translation constant* in MHz and F_{Up} is your chosen uplink frequency. For RS-10/11 and RS-12/13, the F_T value depends on the mode:

Mode A: –116.5 MHz
Mode K: 8.2 MHz
Mode T: 124.7 MHz

Let's use mode A on RS-10 as an example and assume that you've chosen 145.870 MHz as your uplink frequency.

–116.5 + 145.87 = 29.37 MHz

See how easy it is? If you're transmitting on 145.87 MHz, you can expect to find your downlink signal in the vicinity of 29.37 MHz.

watts seems to work well—although I managed to work a station through RS-10 with only 5 watts.

With the wide separation between uplink and downlink frequencies, you'll work the RS satellites in *full duplex*. In other words, you'll be able to hear your own signal on the satellite downlink *while you're transmitting*! When you communicate through the RS satellites for the first time, make sure to keep one hand on your uplink VFO control. It's all due to a pesky little problem known as *Doppler shift*.

The Mystery of the Shifting Signal

Doppler shift is caused by the difference in relative motion between you and another object. As the satellite moves toward you, the signal frequencies in the downlink passband gradually *increase*. As the satellite starts to move away, the frequencies decrease.

When I made my first RS contact on CW, I noticed that I had to adjust my 2-meter transmitter as my downlink signal drifted through the passband. If the satellite is particularly busy, it's possible that one QSO could actually drift into another! As you gain experience with the satellite, you'll discover how to select uplink and downlink frequencies that minimize the chances of a collision.

RS Operating Techniques

Since the satellites are available for only about 10 or 15 minutes, contacts tend to be brief. CW operators congregate in the lower half of the transponder passband while SSB operators occupy the upper half.

If you hear someone calling CQ on SSB, note the downlink frequency and *quickly* tune your 2-meter transmitter accordingly. As you answer the call, adjust your 2-meter transmitter until your voice is clear and audible on the downlink. You can even do this *while he is still calling CQ*. I've heard some SSB operators adjusting their uplink frequency and saying, "Test, test, test..." By using this method they're assured of being on-frequency and ready to respond when the other station stops calling.

Answering a CW call is just as easy. As soon as you copy the call sign, tune your 2-meter transmitter to the proper frequency and start sending a series of *dits*. Listen on the downlink and adjust your transmitter until you hear the tone of your CW signal roughly matching the tone of the station sending CQ.

If you're using a 2-meter FM rig as your CW uplink transmitter, you're probably limited to tuning in 5 kHz steps. This makes it difficult for you to tune onto other stations when they're calling CQ. In this situation it's often best to simply stay in one place and call CQ yourself. (The hams who own the more "agile" radios will come to you!)

The RS satellites also feature CW *Robots*. The robots will call CQ when they are active. To reply, send RS10 DE N1BKE $\overline{\text{AR}}$, for example, at 10 to 30 WPM. If you're on the correct frequency and the robot hears you, it will respond and send a contact number. Reference this number on your QSL card and send it to Box 88, Moscow. See Table 5-2 for robot frequencies.

The future of the RS satellites will depend on the level of funding required to keep them operational. Considering the economic hardships facing the Russian Federation at the time of this writing, there is some cause for concern. With any luck, these satellites will continue functioning, providing many enjoyable contacts for years to come.

The PACSATs

If you enjoy packet operating, you'll love the PACSATs! Several satellites comprise the PACSATs: AMSAT-OSCAR 16, Webersat-OSCAR 18, Lusat-OSCAR 19, Fuji-OSCAR 20, AMSAT-OSCAR 21 and UoSat-OSCAR 22.

OSCAR 20 (also known as FO-20) is essentially an orbiting packet bulletin board. It even uses commands similar to terrestrial bulletin boards. You can pick up and deliver packet mail on OSCAR 20 as well as read the latest general-interest bulletins (see Table 5-3 for frequencies).

OSCARs 16, 19 and 22 are known as "file servers." (OSCARs 16 and 19 are Microsats.) You can access these satellites only by using specialized PACSAT software available from AMSAT. (See the Resources and References Guide for more information.) The AMSAT software permits highly efficient uploading and downloading of files as the satellites pass overhead.

OSCAR 18—known as WO-18, or *Webersat*—is especially fascinating (see Fig 5-2). Another member of the Microsat family,

OSCAR 18 carries an on-board camera and transmits digitized images of the earth, sun and other objects in space. Once again, special software—available from AMSAT—is required to view Webersat images.

Some digital satellites—such as OSCARs 20 and 21—are equipped with linear transponders as well. See Table 5-2.

PACSAT Equipment

Like the other satellites we've discussed so far, the PACSATs *do not* require large antennas and hefty amounts of RF power. (A

Table 5-3

Pacsat Frequencies and Modes

Satellite	*Uplink(s) (MHz)*	*Downlink(s) (MHz)*	*Data Format*
AMSAT-OSCAR 16	145.90 145.92 145.94 145.96	437.025 437.05 2401.10	1200 bit/s PSK AX.25 (4800bit/s, switchable)
DOVE-OSCAR 17	None	145.825 2401.22	Digitized voice with 1200 bit/s AFSK AX.25 (FM) Telemetry
WEBERSAT-OSCAR 18	None	437.075 437.10	1200 bit/s PSK AX.25
LUSAT-OSCAR 19	145.84 145.86 145.88 145.90	437.125 437.15	1200 bit/s PSK AX.25 (4800 bit/s, switchable)
Fuji-OSCAR 20	145.850 145.870 145.890 145.910	435.910	1200 bit/s PSK AX.25
AMSAT-OSCAR 21	435.016 435.155 435.193	145.983	Rudak 2 (other modes and data rates selectable)
UoSat-OSCAR 22	145.900 145.975	435.120	9600 bit/s FSK

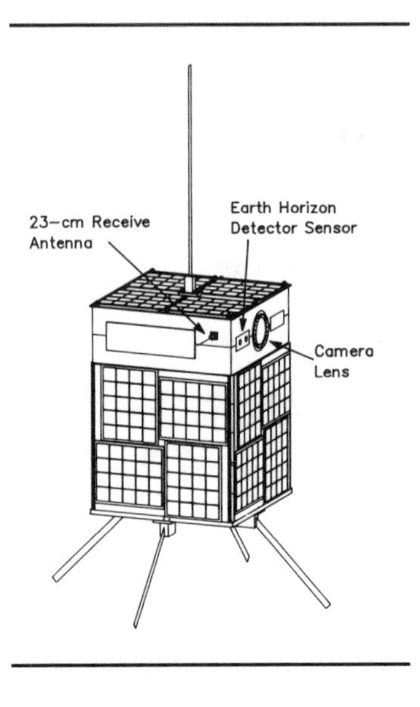

Fig 5-2—An exterior view of Webersat, a Microsat that transmits digitized images of the earth, sun and other objects in space. (*drawing courtesy WD4FAB*)

few watts to an omnidirectional antenna is perfectly adequate.) However, the PACSATs operate in *mode J:* a 2-meter uplink with a 70-cm (437 MHz) downlink. The exception is OSCAR 21, which operates in mode *B:* 70-cm uplink and 2-meter downlink. You may already own a 2-meter FM transmitter suitable for a mode-J uplink, but you'll need a 70-cm SSB receiver for the downlink. Most hams use 70-cm all-mode transceivers for downlink reception. An alternative is to use receive converters to shift a 70-cm signal down to 10 meters for reception on standard HF rigs.

The PACSATs use a variety of data rates and signal formats. Special TNCs are required to communicate with the satellites. See the Resources and References Guide for more information.

Regardless of which TNC you choose, your 70-cm downlink receiver must be capable of automatic frequency control—usually through the UP/DOWN pins on the microphone connector of your 70-cm transceiver or HF radio. The reasoning behind this requirement involves Doppler shifting.

As we've discussed, the frequency drift that takes place on the 10-meter RS downlink can be easily managed through manual adjustments. At 435 MHz, however, it's a totally different story! The higher the downlink frequency, the worse the Doppler shift

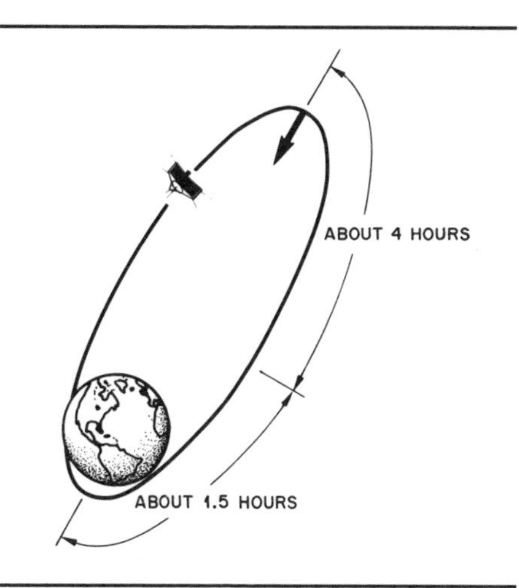

Fig 5-3—The high, elliptical orbit of AMSAT-OSCAR 13 allows it to "see" a large portion of the globe. This means that amateurs separated by vast distances are able to communicate through the satellite. For example, hams in Canada and Argentina can easily enjoy QSOs via AMSAT-OSCAR 13.

becomes. While it's not impossible to compensate manually at these frequencies, it is extremely difficult. Satellite TNCs provide a frequency compensation function which uses the UP/DOWN frequency control (available on many transceiver microphones) to adjust the downlink frequency automatically.

OSCAR 13

Throughout this chapter we've been talking about satellites that travel in low-earth orbits. The advantages of these satellites are obvious: you can access them with relatively low power and very meager antennas. On the other hand, they are available only for brief periods of time and the opportunities for DX are almost nonexistent.

Fortunately, there is a satellite that travels in a high, elliptical orbit, and it's a DX powerhouse: AMSAT-OSCAR 13! OSCAR 13 (or AO-13) is a sophisticated *Phase 3* satellite that incorporates several frequency modes and highly sensitive receivers. It supersedes AMSAT-OSCAR 10, which was still "alive" as of this writing, but damaged and difficult to control.

As you'll see in Fig 5-3, OSCAR 13 has an orbit that acts like a slingshot, shooting it out to an altitude of 30,000 km at its *apogee*. During the high point of its orbit, OSCAR 13 seems to be nearly motionless from our perspective here on earth. While a certain amount of antenna aiming is required, very little additional movement is necessary once the antennas are in their proper positions. From its high vantage point, OSCAR 13 "sees" a great deal of the earth. This opens a window to DX contacts on a regular basis!

The downside to having such a high-altitude orbit is that more transmitted power is needed to access the satellite and a weaker signal is received here on earth. You'll need high-gain, directional antennas to operate OSCAR 13. The more gain you have at the antenna, the less power will be required at the transmitter. You'll need to be able to rotate the 2-meter and 70-cm antennas vertically *and* horizontally (elevation and azimuth).

Mode B (70-cm uplink/2-meter downlink) is the most popular mode on OSCAR 13 and it's the easiest mode for the beginner. (See the frequencies listed in Table 5-2.) A 2-meter SSB/CW receiver is required for the downlink and a similar 70-cm transmitter is necessary for the uplink. Considering the weak signals, a 2-meter mast-mounted preamplifier is also a worthy addition to your OSCAR 13 station.

Like the RS satellites, OSCAR 13 employs a linear transponder. SSB and CW are the modes of choice, although QSOs tend to be longer and more relaxed. With OSCAR 13 you don't have to worry too much about losing the satellite in the middle of your conversation! For more information on operating OSCAR 13, see the *ARRL Operating Manual* or the *Satellite Experimenter's Handbook*.

The Future

Amateur Radio satellites—or commercial satellites with

Amateur Radio capability—seem to appear with increasing frequency. One eagerly awaited satellite is AMSAT-Phase 3*D*. (It doesn't get an "OSCAR" designation until it actually reaches orbit.) Phase 3*D* will be the "new and improved" replacement for OSCAR 13. It will provide even easier access and more user features.

On the distant horizon is Phase 4—the first *geostationary* Amateur Radio satellite. When Phase 4 is in operation, rotatable satellite antennas will no longer be necessary. From our position on earth, the satellite will appear to be perfectly motionless in the sky—a powerful, orbiting repeater accessible 24 hours a day!

So what are you waiting for? Start now by working the easier satellites and move up the ladder as you gain experience. Not only will you enjoy ham radio as you've never enjoyed it before, you'll be more than ready when future dreams become reality!

CHAPTER 6

Smile—You're On Ham Radio!

By Brian Battles, WS1O

magine television without goofy game shows, sappy sitcoms or hour-long "infomercials." No more humdrum talk shows, dreary newscasts or monotonous cartoons. TV without commercials? Live broadcasts from studios and remote locations around town, in the air or from a nearby mountaintop? Television programs produced, transmitted and hosted by your friends and neighbors? This is the world of amateur television (ATV)!

ATV can be operated by anyone with a ham radio license. Novices, for example, have ATV privileges from 1270 to 1295 MHz. Hams licensed at the Technician class or higher enjoy the most flexibility. They have full access to all amateur frequencies above 50 MHz, and most ATV is in the 420-440 and 902-928 MHz bands. If you're able to take advantage of these privileges, you can have a great time with television.

It's Not the "Boob Tube"

Think of the fun you have with voice, packet, CW and other modes. Now imagine adding live, moving pictures with sound!

WB6BAP installs a 10-GHz ATV link in preparation for the annual Rose Parade in Pasadena, California. The video transmissions allow parade officials to easily monitor the progress of the event. (*photo by Bill Brown, WB8ELK*)

You can send video shots of yourself, your shack, your family, interesting QSL cards (yours and ones you've collected) and scenes from around town. You can also transmit video of your club's Field Day operation, weather data or just about anything you can think of (as long as it's not commercial or obscene, of course!).

Local public service agencies and government emergency preparedness offices are enthusiastic about having experienced hams provide television coverage of community events, drills and disasters. You can take your portable ATV station aboard a police or National Guard helicopter to survey the extent of forest fires, floods and storm damage. Parades, walkathons, bicycle races and other outdoor activities can be supported by skillful placement of mobile ATV units sending video images to command posts. Trained weather observers can aid National Weather Service officials by letting them see developing storms or tornadoes firsthand.

Avid ATVers enjoy an extraordinary range of operating exploits. You can mount a simple camera, transmitter and

Students at Southeastern Community College in Whiteville, North Carolina, prepare to launch a sophisticated ATV-equipped rocket. A balloon will carry the rocket to 90,000 feet where a command signal from the ground will ignite the rocket engine. (*photo by Bill Brown, WB8ELK*)

associated electronics in a compact package and launch it as payload a helium balloon. Experimenters have amassed thrilling videotape recorded from the received signals of ATV-equipped balloons at the edge of space, more than 100,000 feet up! The view from an altitude of 20+ miles is breathtaking. Closer to the ground, hams combine hobbies by controlling their radio-controlled aircraft, boats and cars with a "pilot's-eye view" monitored from a camera mounted in their miniature craft. Amateur rocketeers mount ATV equipment inside model rocket nosecones for a rapid ride up to 1000 feet or more, watching the world recede beneath their "eye in the sky."

Get a Glimpse of ATV

In any given month, there are at least one or two ATV

Table 6-1

Popular 70-cm ATV Frequencies (MHz)

Video	*Audio*	*Use*
421.25	425.75	repeater outputs
426.25	430.75	repeater inputs or outputs
434.00*	438.50	simplex and repeater inputs
439.25*	443.75	simplex, repeater inputs and outputs

*These are the most widely used ATV frequencies in the US.

experiments conducted somewhere in the US. Perhaps you'll be close enough to watch the fun! From spring through autumn, ATV fans launch high-altitude balloons and rockets that have transmissions you can receive on your standard cable-ready television or video tape recorder. Many portable LCD televisions can tune ATV frequencies directly. To watch ATV with your own receiver, all you have to do is tune to the appropriate "channel" and point an antenna in the right direction (see Table 6-2).

Packet bulletins and HF nets are excellent sources of information about major operating events. Regional ATV nets meet Tuesday evenings at 8 PM on 3871 kHz and Sundays at 11 AM on 7243 kHz. Check into your nearby packet bulletin board (PBBS) and scan through the mail for ATV announcements. If you don't find any,

Table 6-2

Cable TV Channels by Radio Frequency

Channel	*MHz*
57	421.25
58	427.25
59	433.75
60	439.25

post one yourself, asking whether anyone in the region is an active ATVer. You may be surprised at the responses you receive.

Getting Started

At first glance, ATV seems like a costly activity. If you use broadcast television stations as your guide, you're looking at a substantial investment for transmitting gear, video monitors, cameras, sync generators and so on, right?

Wrong!

Unless you choose to set up the ultimate station, ATV is very affordable. If you happen to have a typical home video camera or camcorder and a basic TV set (black and white is fine), you can

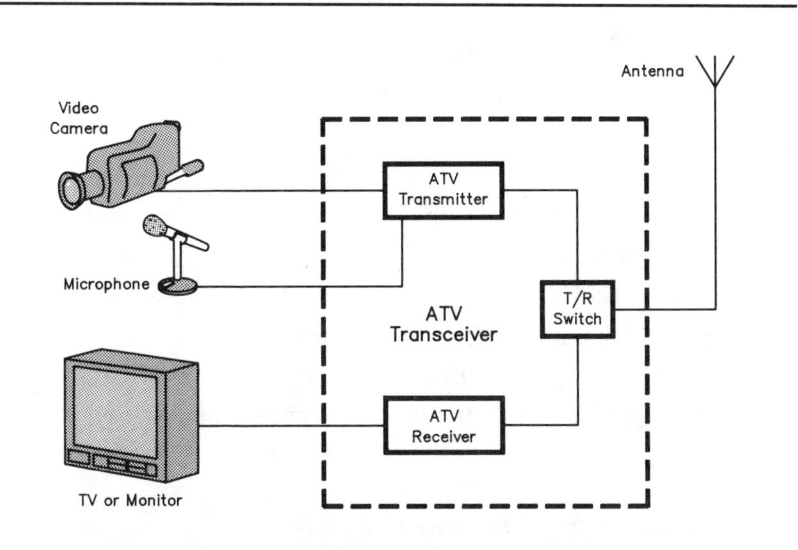

A typical ATV station consists of a video camera (with a self-contained or separate microphone), a transmitter, a receiver and a standard TV or a video monitor. Early ATV stations used separate transmitters and receivers. Manual or remote-controlled transmit/receive switches connected the main antenna to one unit or the other. Modern ATV stations now feature easy-to-use *transceivers* with the transmitter, receiver and switch in a single cabinet.

If You Don't Own a Video Camera...

Consider buying a camcorder. Prices have been falling in recent years and some camcorders are showing up on the used equipment market. Don't forget that a camcorder is one of the few items in your ham shack that has plenty of value for family activities. This fact may make it easier to justify the purchase.

If you can't afford a camcorder, look for a used *closed circuit* (CCTV) camera. Most of these cameras are black and white and their resolution (image detail) isn't the greatest, but they'll be sufficient to get you started. Used CCTV cameras are often found at hamfest flea markets, sometimes for as little as $30. If you buy a used CCTV camera, be prepared to do a little work. You may have to build a special power supply, or troubleshoot a few problems. Even so, if you purchase a camera that's in decent shape, you'll be on ATV at a bargain price!

purchase an ATV transceiver for about $400. Add perhaps $75 for a reasonable antenna and a few dollars for cables and miscellaneous odds and ends. You can get your TV station on the air for less than $500—and that's if you insist on new equipment! You can build your own from scratch or from a kit, or buy inexpensive used gear. With a little resourcefulness, you may be surprised to find yourself the proud owner of a television station for an investment of $100 or less.

Antennas, Amplifiers and Preamplifiers

Don't worry about putting up a monster antenna system for ATV. You're not building a world-class contest station for 160-10 meters, so set aside your concern for having to erect a 200-foot tower with 50-foot Yagis and aircraft engine-sized rotators.

All amateur fast-scan television (FSTV) is on UHF and microwaves, so high-gain antennas aren't too much trouble to build. A high-gain multielement Yagi for 434 MHz is only two or three feet long and gives you a big boost on transmitting and receiving.

If there is an ATV repeater in your area, a simple omnidirectional antenna may be adequate. Check the *ARRL Repeater Directory* and talk to ATV operators in your town. Find out where the repeater is located and how sensitive it is. Assuming that you can receive a strong signal from the repeater—and assuming the repeater can hear your signal as well—a UHF ground plane antenna may be all you'll need.

When choosing your ATV antenna, don't forget about *polarization*. Most antennas used in ATV applications are either horizontally or vertically polarized. Yagi antennas, for example, can be horizontal or vertical depending on how you mount them

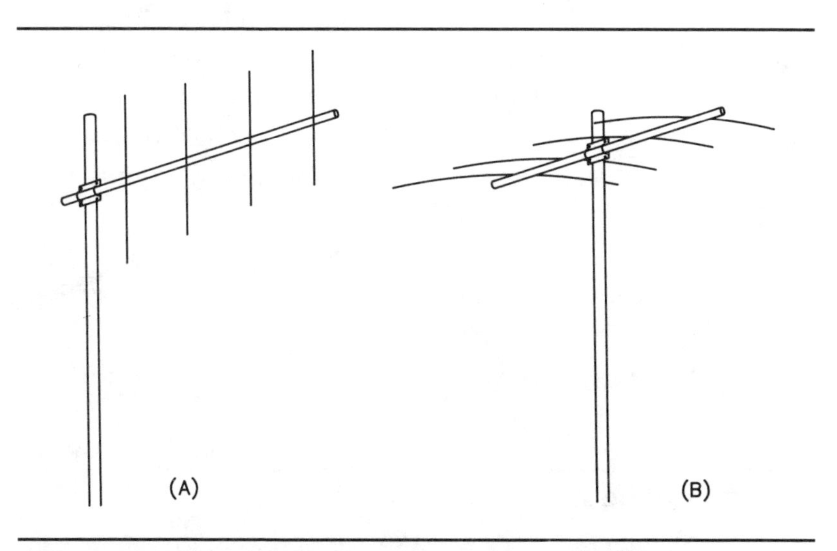

Fig 6-1—The Yagi antenna shown at A is mounted in the vertically polarized position with its elements perpendicular to the ground. This is the best polarization for using ATV repeaters. The Yagi shown at B is horizontally polarized with its elements parallel to the ground. Horizontal polarization is ideal for direct (simplex) contacts and ATV DXing.

(see Fig 6-1). Ground planes, on the other hand, are vertically polarized. Using the proper polarization is important. If you're vertically polarized and the station you're talking to is horizontally polarized, you'll both experience a noticeable loss in received signal strength.

So which antenna polarity should you use? It all depends on the ATV activity in your area. If you intend to operate through an ATV repeater, vertical polarization is best since most repeaters use vertically polarized antennas. For direct contacts—or ATV DXing—horizontal polarization is the standard. Some antenna manufacturers allow you to enjoy the best of both worlds with dual-polarization antennas. These are basically two Yagi antennas mounted on the same boom. One is in the horizontal position and the other is vertical. An antenna switch (often remotely controlled) is used to select one antenna or the other.

In marginal areas, a receiver preamp mounted at the antenna can be a big help (installing a preamp is probably a good idea no matter where you live). If you have a powerful TV broadcast transmitter nearby, you might have to add some high- or low-pass filters to keep the commercial station from overwhelming your receiver.

If you aren't within a dozen miles or so of a local ATV repeater, think about adding a power amplifier to your station. Most typical ATV transmitters put out 1 to 5 watts, and boosting that to 50 or even 100 watts could bring your signal up from barely copyable to sharp and clear!

You're on the Air!

How do you call "CQ" on ATV? If you want to be informal, just sit down in front of the camera and make yourself comfortable. Activate your transmitter and say, for example, "This is KY1T in Newington, Connecticut. Anyone around?" Switch back to the receive mode and wait for a response. If no one replies, choose

The W5KPZ ATV repeater in Tyler, Texas. With an ATV repeater nearby, even a low-power station with a marginal antenna can extend its coverage over a wide area. When this photograph was taken, AA5BY was transmitting through the system. His video signal can be seen in the repeater monitor. ***(photo by Bill Brown, WB8ELK)***

another frequency (or repeater) and try again.

Another approach is to broadcast an image of your call sign for several seconds while announcing your invitation. Your call sign should be large enough to fill the entire screen. Bold, black letters on a white background will make it easier to read at greater distances (black lettering on white cardboard will do nicely). Computers or character generators can also be used to create attractive ID screens. Let your imagination be your guide!

You can establish contact with another station on a single frequency (simplex), or use an ATV repeater with separate input and output frequencies (duplex). The same operating procedures apply in either case. Some ATV repeaters transmit a continuous *beacon* (usually composed of its call sign and some graphics) to help you find it. Other ATV repeaters transmit their identifications at regular intervals (on the hour and half-hour, for example). When you transmit, the beacon will disappear and your video signal will be relayed through the system. When you stop transmitting, the beacon will return.

When you finally establish contact, the first order of business is to exchange signal reports so that each of you will know how well your signals are being received. In ATV we use the "P" system to describe the amount of noise (or "snow") in the picture. A P5 picture, for example, is excellent with no visible noise. A P1 picture is very weak with a great deal of noise (see Fig 6-2).

After that, anything goes! You can talk about whatever comes to mind. Unlike other Amateur Radio modes, the person at the other end will be able to see your gestures and facial expressions as you speak. This gives ATV a unique personal dimension that is difficult to achieve on SSB, CW, FM or packet.

The 2-meter Connection

In many areas of the country, it's a popular practice to use a 2-meter FM simplex frequency—or a repeater—for local ATV coordination. This allows ATVers to make critical adjustments *while transmitting*. Since mutual interference between the 2-meter and 420-MHz equipment is minimal, ATVers can get instant comments on their signal quality and make additional adjustments as necessary.

"WB8QVC from WB8SVN. Your signal is getting worse. You turned your beam too far."

"There. The antenna should be pointing directly at the repeater. How's that?"

"Much better, Tom. I can see you and the chair you're sitting in, but everything else is pretty dark."

UNITED STATES ATV SOCIETY
AMATEUR RADIO FAST SCAN TELEVISION
VIDEO PICTURE STANDARDS

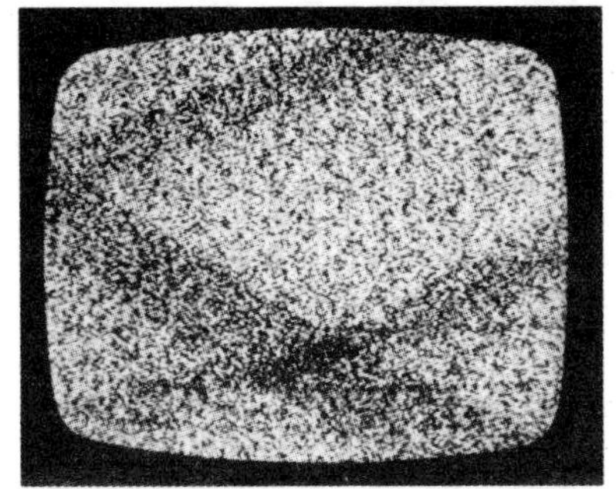

P0 Total Noise Visable. No picture at all or detectable Video Sync Bars.

P1 High noise visable. Weak picture.

P2 High noise visable. Fair picture. Fair detail.

P3 Noise Visable. Strong picture.Recognizable detail.

P4 Slight noise visable. Very strong picture. Good detail.

P5 No noise visable. Closed circuit picture. Excellent detail.

Fig 6-2—The ATV picture quality reporting system. A P5 picture is very clear and easy to see. On the other hand, a P0 signal cannot be seen at all.

"That's a lighting problem. Let me switch on my desk lamp."

"Nice! Now I can see everything. Say, that's an impressive movie poster you have on your wall. Where did you get it?"

Many enthusiasts will call CQ on 2-meter FM and 70-cm ATV simultaneously. This technique is effective in attracting the attention of local operators who may be monitoring 2 meters while their ATV equipment is inactive. If ATV coordination is taking place on 2 meters in your area, try to find the frequency. (144.34 MHz is popular.) By listening to the chatter you'll get a pretty good idea of how many ATV stations are on the air, where they are located and when they are active.

ATV DXing

There is more to ATV than chatting with your local friends. You'll also have opportunities to enjoy QSOs with ATVers hundreds of miles away. It doesn't happen every day, but when it does, ATV DXing offers genuine excitement!

If you plan to work ATV DX via direct, single-frequency contacts, you'll need to invest in the best equipment you can afford. Antenna-mounted preamplifiers are mandatory along with RF power amplifiers and low-loss coaxial cable or hardline. Some

John, KDØLO, is smiling for a good reason! He's an accomplished ATV DXer, working stations over 400 miles away from his home in St Louis, Missouri. (*photo by Bill Brown, WB8ELK*)

ATV DXers use a single, high-gain Yagi antenna (horizontally polarized). Others use several Yagis working together in a carefully designed assembly fed by the primary coaxial line. This is often called an *array*. Depending on its construction, an array is capable of very high performance in DX applications.

You can also work DX through your local ATV repeater. When the band is open, many ATVers will attempt to reach repeaters in distant locations. Don't be surprised if you call CQ on your repeater one evening and receive a reply from an ATV DXer!

The best times for ATV DX are around sunrise and right after sunset. If your family TV is connected to an outside antenna, tune through the commercial UHF Channels. If you begin to see distant stations, especially on channels 14 through 30, there may be a band opening in progress!

Television Stardom Awaits You

By the time you've reached this point in the chapter, there must be at least a spark of interest in your heart for ATV. Start exploring the activity in your area. Drag out that VCR or cable-ready TV tonight and see what you can find. You can also buy or

ATV pioneer Don Miller, W9NTP, in his hamshack in Waldron, Indiana. (*photo by Bill Brown, WB8ELK*)

build ATV *downconverters* that will receive ATV transmissions and convert them to Channel 3 or 4 for viewing on your television. If you're using the family TV antenna for your explorations, don't be surprised if the ATV signals are not as strong as you'd expect. When you finally put up a proper ATV antenna system, you'll see a vast improvement.

Check the Resources and References Guide at the back of this book for more information and the addresses of ATV equipment manufacturers. Once you put together an ATV station and get on the air, you'll be hooked forever!

CHAPTER 7

Awards

By Mark Wilson, AA2Z

In Chapter 4, we discussed SSB/CW operation, band openings and DXing. Many hams get on during band openings for the thrill of working someone far away—after all, DXing is a basic human instinct. More often than not, though, there's another motivating factor—awards chasing!

Are you a collector? For hams on VHF, a favorite pastime is collecting contacts with different geographic areas—grid squares,

If you want to attract a lot of attention during the next VHF contest, find a rare grid square and set up your station there. Tony, KC4AUF, had the best of both worlds during the 1991 June VHF QSO Party. Grid FM26 is not only rare, it's near a popular vacation spot: Virginia Beach, Virginia!

The K1TR group operated from the cloud-shrouded summit of Mt Washington in New Hampshire (rare grid FN44) during the 1991 September VHF QSO Party. This UHF/microwave array snagged numerous contacts from an elevation of over 6,000 feet!

states, continents and countries. Collect enough contacts, and you can earn an award. Most award programs work the same way: Contact the required stations, collect QSL cards from them to confirm the contacts, and present the cards for verification. Once the cards are verified, you receive a handsome certificate.

There are a lot of good reasons for collecting awards. Getting a certificate to frame is really an anticlimax. For most people, the fun is in working toward and achieving a goal. It's honing your operating skills so you can work that oh-so-weak station in a rare area. It's learning the ins-and-outs of VHF propagation modes, and the techniques necessary to exploit those modes to work far-off stations. It's gaining some technical savvy and improving your station and antennas to hear—and be heard by—stations hundreds or thousands of miles away. It's swapping stories with local hams who are chasing awards, too.

Choose the Awards You Want to Earn!

The ARRL offers four awards of interest to VHFers:

- ❑ *VHF/UHF Century Club (VUCC)* for working a specified number of grid squares.
- ❑ *Worked All States (WAS)* for working all 50 states.
- ❑ *Worked All Continents (WAC)* for working all six continents.
- ❑ *DX Century Club (DXCC)* for working 100 different countries.

Two special-interest clubs offer awards as well:

- ❑ *VHF/UHF County Award (VUCA)* is offered by Side Winders on Two (SWOT) for working counties on VHF.
- ❑ *DX Decade Club* and other awards are offered by the Six Meter International Radio Klub (SMIRK) for working DX countries on 6 meters.

SWOT and SMIRK also offer awards for working members of their organizations. See the Resources and References Guide for more information on SWOT and SMIRK.

VHF/UHF Century Club (VUCC)

The VHF/UHF Century Club is specifically tailored around operation on the VHF and UHF bands. It's the award that most VHFers shoot for, and earning one is quite an accomplishment! VUCC is available to ARRL members in the US and Canada, and to other amateurs worldwide.

VUCC is awarded for contacting stations in a minimum number of grid squares (see Chapter 4 for a discussion of grid squares). Awards are issued separately for each band, and the minimum number of grid squares needed to make the grade varies from band to band. It's important to note that contacts from different bands cannot be combined for credit, and that crossband contacts (receiving on one band, transmitting on another) are not permitted. Once you've earned the basic award, you can work for endorsement stickers for working additional grid squares. Table 1 shows the minimum number of grids required for an initial award and endorsements for each band.

One of the most popular awards for VHF operators: The VHF/UHF Century Club (VUCC).

The VHF/UHF Century Club has a few simple rules. You can work stations on any legal mode (FM, SSB, CW, ATV, and so on), but contacts made through repeaters or other power relay devices don't count. There are no special endorsements such as "FM only" or "SSB only," so go ahead and use whatever modes you can to work new grids. Contacts with fixed, portable or mobile stations count, too. You can work stations in boats (maritime mobile), but *not* in airplanes (aeronautical mobile).

For awards through 1296 MHz, you must make all your contacts from a location or locations within the same grid square, or from locations in different grid squares no more than 50 miles apart. This means that if you work some new grids while you're operating portable from the local hilltop one warm summer evening, you can add them to the grid totals you've racked up from home. For the bands above 1296 MHz, however, you've got to make all contacts from a single location (all your equipment has to be placed within a 300-meter circle).

There's a separate award for working grid squares via amateur satellites. In this case, contacts made through any amateur satellites can be grouped together for credit. The prohibitions against crossband and repeater (power relay) contacts are waived, of course, because of the mechanics of amateur satellite communications.

A favorite activity, especially during VHF contests, is activating rare grid squares. Many grids, especially in rural areas, have few, if any, active VHF or UHF operators. It's not uncommon for someone to load up the car or truck with gear and antennas and drive to one or more inactive grids during a weekend. A few enterprising souls even take boats to all-water grids just off the coast. If you listen regularly on the bands, you'll hear news of these operations—they attract a lot of attention!

Table 1

The Number of Grid Squares You'll Need to Earn Your ARRL VHF/UHF Century Club Award

Band	*Grids Needed for Initial Award*	*Grids Needed for Endorsement*
50 MHz	100	25
144 MHz	100	25
222 MHz	50	10
420 MHz	50	10
902 MHz	25	5
1.2 GHz	25	5
2.3 GHz	10	5
3.4 GHz	5	5
5.7 GHz	5	5
10 GHz	5	5
24 GHz	5	5
47 GHz	5	5
Laser	5	5
Satellite	100	25

One question inevitably arises from portable operation: "What if I operate from the intersection of *four* grid squares? Will the contacts count for all four grids?" The answer is, you *can* operate from more than one grid simultaneously, but it's not easy. Your station must be arranged so that pieces of it are physically present in the different grid squares. You must also know the *precise* location of the boundary, either from markers permanently in place or from the work of a professional surveyor.

Now that you know the rules, get on the air and start making contacts! It's helpful to get some copies of the ARRL grid square map of North America (see the Resources and References Guide) so you can track your progress. There's a lot of satisfaction in coloring in new grids as you go.

Be prepared to learn about and look for sporadic-E, tropo, aurora and other kinds of VHF/UHF DX propagation in your quest for new grids. Although 100 grids on 6 or 2 meters seems like a lot, there is enough activity to keep things moving. Contests are a good time to look for new ones. The top stations often work 100 or more grids on 6 meters during a good June VHF contest, and there is often activity from rare grids. To give you an idea of what's possible, several stations have worked 600-700 grids on 6 meters and more than 400 grids on 2 meters since VUCC started on January 1, 1983.

After you've worked the required contacts and collected the QSL cards, it's time to apply for the certificate. Write to ARRL HQ and request the paperwork. Make sure to include a self-addressed, stamped envelope (SASE). You'll need forms MSD-259, MSD-260 and MSD-261. These are the application forms and a complete set of rules, including the application procedures. Your QSL cards and completed application must be verified by an approved VHF Awards Manager (many clubs have one). If you don't know the name of the closest VHF Awards Manager, ask for this information from ARRL HQ when you request the forms.

Worked All States (WAS)

Worked All States (WAS) requires contacts with all 50 states. Even though many hams think of WAS as an HF award, several hundred amateurs have earned this award on the VHF bands. The League offers separate awards for working 'em all on 50, 144, 222 and 432 MHz, as well as for amateur satellite operation. WAS is open to ARRL members in the US and Canada, and to other amateurs worldwide.

Although many VHF hams collect new states, working all 50 on 6 meters or above requires a real commitment. Six meters is where most of the action is; more than a hundred awards have been issued for that band. On the higher bands, it's often necessary to use moonbounce (see Chapter 4). There's even an endorsement sticker available for moonbounce work. A handful of die-hards have worked 48 states on 2 meters without moonbounce, but most hams can eventually work 35 to 45 states on terrestrial modes. Obviously a lot depends on which part of the country you live in!

If you're active on

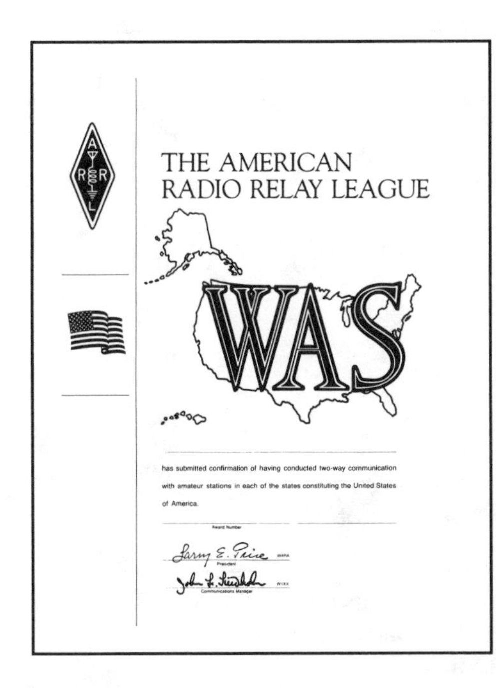

If you make a contact with an amateur in every state, you'll qualify for the Worked All States award (WAS).

amateur satellites, you won't have to wait for special propagation or build a moonbounce station to work your 50 states. Many hams have achieved WAS using a number of different "birds."

Like VUCC, cards and applications forms can be verified locally. Check with your local club or ask the Awards Branch at ARRL HQ for the name of the HF Awards Manager nearest you. Complete rules (MCS-264) and application forms (MCS-217) are available from ARRL HQ for an SASE.

Worked All Continents

Worked All Continents (WAC) is not for everyone: You have to work six continents—Africa, Asia, Europe, North America, Oceania and South America—to earn it. Sponsored by the International Amateur Radio Union (IARU), this award is administered by ARRL and is open to everyone. The basic WAC

Bagging your Worked All Continents award is a major achievement on VHF! Many hams use satellites and moonbounce in pursuit of their WAC.

certificate is offered with endorsements for 50, 144 and 432 MHz, as well as for any higher band. A separate award is available for working all continents via amateur satellites.

With all the interesting DX stations to work on amateur satellites, WAC is something to shoot for. Some knowledge of the various satellites' orbits is helpful to work the continents that are farthest from you.

On 6 meters, you'll need the assistance of F-layer propagation to work all six continents. It's a challenge, even at the peak of the sunspot cycle when 6-meter DX openings happen regularly.

Rag Chewers' Club—The Easiest Award of All!

There's one award that you can qualify for right away. Nearly every awards hunter starts off with a Rag Chewers' Club (RCC) certificate. RCC encourages friendly, meaningful contacts, rather than "hello-goodbye" QSOs. To earn RCC, start a nice leisurely chat with someone on the local FM simplex frequency. Simply enjoy the conversation for a half hour and report the contact to the Awards Branch at ARRL HQ (please enclose an SASE). Your Rag Chewers' Club certificate will be sent by return mail. Contacts on *any* band or mode are eligible.

Side Winders On Two (SWOT) sponsor the VHF/UHF County Award (VUCA) to encourage activity above the 6-meter band. By working amateurs in 100 US counties, you'll qualify for the initial 6- or 2-meter awards (see Table 2).

WAC is available to moonbounce operators on 144, 432 and perhaps 1296 MHz. It's not possible on 222 or 902 MHz because these bands are not available for amateur operation outside North and South America. Stations from all six continents have been active on moonbounce in recent years.

Complete rules (MCS-264) and application forms (MCS-217) are available from ARRL HQ for an SASE.

DX Century Club

If you're looking for a *real* challenge, working 100 countries for the DX Century Club (DXCC) will test your operating skill, technical ability and patience. DXCC awards are available for 6 meters, 2 meters and satellite operation. Several dozen stations qualified for 6-meter DXCC during the sunspot peak of the late

1980s and early 1990s, after years of hard work. So far, only one station—with a monstrous moonbounce antenna—has claimed DXCC on 2 meters. Satellite DXCC, somewhat easier to earn than a terrestrial award, is a challenge as well. DXCC is open to ARRL members in the US and Canada, and to other amateurs worldwide. Complete rules and application forms are available from ARRL HQ for an SASE.

VHF/UHF County Award (VUCA)

Side Winders on Two (SWOT) sponsors the VHF/UHF County Award (VUCA) to encourage VHF/UHF activity and to recognize operating achievements on the frequencies above 50 MHz. VUCA certificates for each VHF/UHF band are awarded for contacts with a minimum number of counties per band. Once you've earned the basic award, you can work toward endorsements for additional counties. See Table 2. VUCA is open to all licensed hams, and contacts made after January 1, 1983, count.

There are more than 3000 counties and independent cities in the US that qualify for this award. These counties are the same as those for *CQ* Magazine's Counties Award. The *CQ USA County*

Table 2

The Number of Counties You'll Need to Qualify for Your SWOT VHF/UHF County Award

Band	*Counties Needed for Initial Award*	*Counties Needed for Endorsement*
50 MHz	100	50
144 MHz	100	50
222 MHz	50	10
432 MHz	50	10
903 MHz	25	5
1.2 GHz	25	5
2.4 GHz & up	5	5

Award Record Book is the best way to keep track of counties worked toward the award. You'll eventually need to record your contacts in one of these books to apply for the award, so why not get one right away? The official county list is also published in *The ARRL Operating Manual* (see the Resources and References Guide).

VUCA awards are numbered separately for each band and may be endorsed for CW, SSB, moonbounce, satellite, ATV, packet, FM simplex, mobile or QRP operation. Contacts via terrestrial repeaters don't count.

When you've collected enough contacts to apply for VUCA, be sure they're recorded accurately in your *Record Book*. The QSL cards must show the county or other identifiable location (city, town, latitude/longitude, for example) for positive county determination. Then you need to get two licensed amateurs to examine your cards and sign the certification form in the *Record Book* to verify that you have, indeed, qualified. The *Record Book* becomes your official SWOT VUCA record and won't be returned.

For USA stations, all contacts must be made from one county or any county immediately adjoining it. Mobiles operating on a county line may be credited for no more than two counties at a time.

Fees: $4 for SWOT members and overseas stations, $5 for USA non-SWOT members, for each certificate. Certificate endorsement, $1 and an SASE.

Send applications and make checks payable to L. G. Parsons, W5AL, SWOT VUCA Awards Manager, 3316 Edenburg Dr, Amarillo, TX 79106, tel 806-352-0835.

SWOT Worked Members Awards

Awards are available for working members of SWOT. The initial award is for working ten members. Successive awards are

Contact 1,000 members of the Six Meter International Radio Klub (SMIRK) and you'll earn the 1,000 SMIRK award.

for working members in levels of 25 (25, 50, 75, 100, and so on). Send $1 and an SASE per award to Jerome Doerrie, K5IS, SWOT Membership Awards Manager, Rt 2 Box 72, Booker, TX 79005, tel 806-658-2264.

SMIRK DX and Membership Awards

The Six Meter International Radio Klub (SMIRK) offers a series of awards for working DX stations on 6 meters. The *DX Decade Club* is earned for contacting 10 DXCC countries. There are endorsements for each additional 10 countries. Even in years when there is no worldwide F-layer DX to work on 6 meters, you should be able to earn this award by working DX on sporadic E.

As your country totals grow, you can earn the *50 Country Club Award* for contacting 50 DXCC countries and the *100 Country Club Award* for contacting 100 DXCC countries.

SMIRK also offers awards for working club members. The

basic award, *100 SMIRK* is available for working 100 members. Advanced awards are offered for working 250, 500 and 1000 members.

Certification for all awards, except the 50 and 100 Country Club awards, is self certification. For the 50 and 100 Country Club awards, photocopies of the QSL cards are required.

Fees: $3, for 1000 SMIRK and DXDC; $5 for 50 and 100 Country Club. 100, 250 and 500 SMIRK seals are $1 each.

Contact Don Abell, KC5TK, SMIRK Awards Manager, 6821 West Ave, San Antonio, TX 78213, tel 512-349-7234, to apply for these awards.

The Challenge and Reward

Some of the awards outlined here are relatively easy to achieve. Others take years of patience, operating skill and an investment in station hardware. There are awards, though, to suit any level of station hardware and operating skill. All have one thing in common: They'll encourage you to get on the air and to learn new things. Collecting contacts toward these awards can only add to your enjoyment of Amateur Radio, so jump in and enjoy!

CHAPTER 8

Transmitter Hunting

By Jim Kearman, KR1S

o matter what area of Amateur Radio interests you, you'll find something to enjoy in transmitter hunting. Aside from the obvious competitive aspects, you can build your own equipment and antennas, assist in public-service activities and have a lot of outdoor fun. Better yet, your family can join in, too. You don't need a ham license to operate a receiver!

Getting Started

Most sport transmitter hunting (sometimes abbreviated "T-hunting") is done on the 2-meter FM band. You can get started with just a portable 2-meter receiver that has a signal strength meter, or *S-meter*. You *can* use a receiver without a meter, but you'll have to rely on your ears to tell you whether the signal is getting stronger or weaker. A meter is a more precise indicator.

One item you'll definitely need is an *attenuator* between the antenna and rig. You can build a simple attenuator from plans in *The ARRL Handbook*. An attenuator is used to deliberately reduce the received signal strength as you get closer to the transmitter. If the signal becomes too strong, you won't be able to pinpoint the source.

You'll also need a directional antenna, but you can build one easily, from plans in *The ARRL Antenna Book*. Like any sport, serious players have more exotic equipment. Like you, most of them started out with simple receivers and antennas.

Sport Hunting

Recreational transmitter hunts are usually organized by clubs. Most hunts are open to non-members. While the participants gather at a local landmark, the transmitter operator is heading into the field to hide. The transmitter may be hidden nearby or far away. Some clubs hide the transmitter over 100 miles away. Their hunts take all day and burn lots of gas! A favorite trick of transmitter hiders though, is to plant a low-power transmitter near the starting location. They position the antenna so the signal reflects off hills or mountains, so it *seems* like the transmitter is 100 miles away! The hidden transmitter is sometimes called the "fox," and some hiders are pretty foxy!

The transmitter comes on the air at a prearranged time. Hunters swing their antennas to take their first bearings. Then everyone jumps into their cars and the chase is on. (Of course, the entire hunt can take place on foot. This type of hunt is an ideal

T-hunters in Albuquerque, New Mexico, lined up at the starting point. *(KØOV photo)*

activity for a club picnic.) Hunting is easier and more fun when done with another person. While one drives, the other checks the maps, swings the beam and listens to hints from the transmitter operator (not all transmitter operators cooperate by giving hints).

Dedicated transmitter hunters are able to take bearings while in motion. Some have even cut holes in their car roofs, through which rotating masts are placed during hunting season. Other operators use *Doppler* direction-finding gear with active antennas connected to indicators in the car. Such installations make transmitter hunting almost luxurious, but you can still find the transmitter with less-complicated gear.

Once you have a rough idea of the area in which the transmitter is operating, you drive as close as you can and take more bearings. (For example, is the fox south or southeast from your present position? Check and make sure!) Each set of bearings should bring you closer to the transmitter. Here's where having two or more people on your team pays off. As you get closer to the transmitter a real traffic jam of hunters will develop. The driver will be too busy avoiding other hunters to do much more than steer!

When you get very close to the transmitter, taking a bearing from a second position lets you *triangulate* to get a better idea of its location (Fig 8-1). If the transmitter is fairly close, the two initial locations from which you take bearings need be only a few hundred yards apart. If the transmitter is far away, it's better to take the initial bearings from more distant locations. The terrain over which the hunt takes place determines the necessary separation, as does the type of antenna you're using.

The Game's Afoot!

Transmitter operators aren't known for making life easy for hunters. In most hunts you'll have to leave your car and walk the rest of the way. You can bet the terrain will be challenging. As

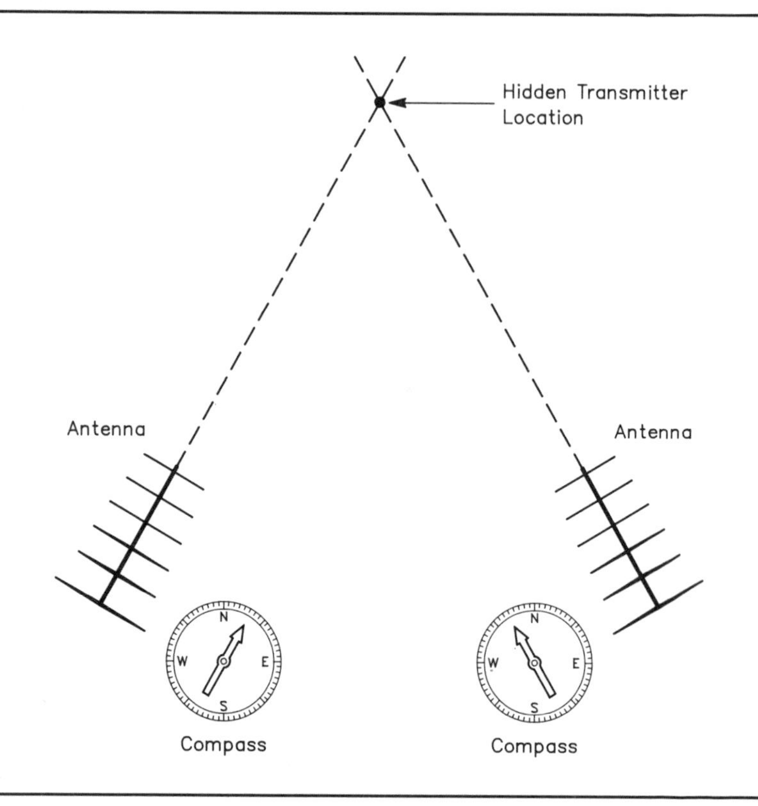

Fig 8-1—When you're close to a hidden transmitter, you can use *triangulation* to pinpoint its location. Take two bearings from locations a few hundred feet apart and draw lines representing the bearings on your map. The transmitter is located very near the point where the two lines intersect.

you get closer to the transmitter, you may find the signal gets so strong that it's hard to determine its peak strength. Using the null off the sides or back of your beam may help. Here's where having an attenuator between the antenna and receiver really comes in handy. You'll also want to make a smaller antenna for hand-held use. When you're extremely close to the transmitter, a simple field-strength meter can take the place of your receiver.

As you finally close in on your target, don't be surprised to find that several others have already beaten you to the spot. Transmitter hunting is like running a marathon: at the beginning, it's important to finish, not win. With experience you'll develop techniques to help you finish sooner.

Games Transmitters Play

If the world was perfectly flat, it would be relatively simple to find any hidden transmitter. Hills, mountains and buildings are good reflectors of VHF signals; so transmitter hunting becomes a real challenge when they're in the vicinity. Fig 8-2 shows a typical case. The hidden transmitter is down in a valley,

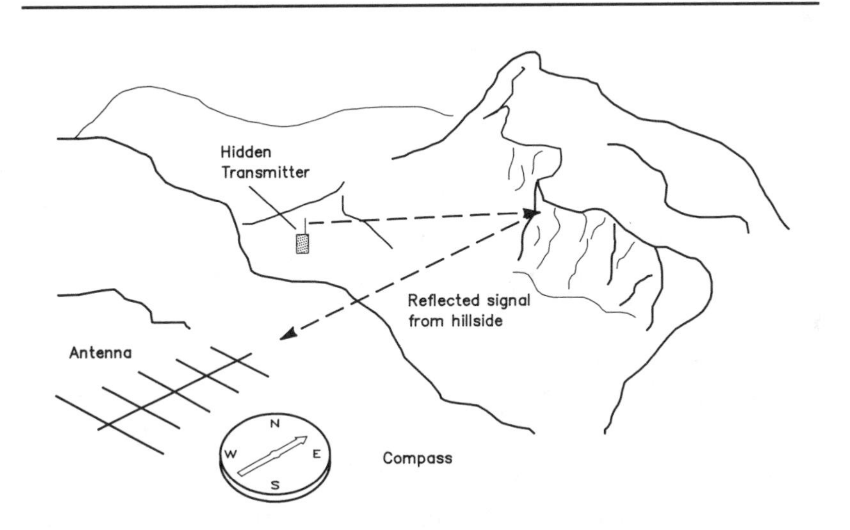

Fig 8-2—A transmitter hidden in a valley can cause misleading beam headings. In this example, the transmitter antenna is a beam, pointed at the mountain. This tricks many hunters into believing that the signal is coming directly from the mountain. With experience you'll be able to detect and identify the weaker signal coming from the transmitter itself. A similar effect occurs when the transmitter is hidden among buildings or other reflective surfaces, like railroad cars or truck trailers.

Jason Pelaez, N8NDQ, buries a 100-milliwatt transmitter for the Dayton (Ohio) Amateur Radio Association Winter Foxhunt. ***(AH2AR photo)***

connected to a highly directive beam antenna pointed at the mountain. From our observation point we detect only the reflection off the mountain. Carefully swinging our antenna *may* detect the direct signal from the back of the beam—if our antenna is directive enough. Before we started climbing that bramble-covered mountain, we'd want to take a few bearings from its base, just to be sure!

Urban transmitter hunts are equally challenging, as a cluster of steel-framed buildings can scatter the transmitter's signal in dozens of directions. The problem gets worse at higher frequencies. If you're hunting in the city, remember that the transmitter need not be at ground level. When you're close to a transmitter hidden a few floors up in a building sandwiched between others, dogged persistence pays off. Use your best directive antenna to "paint" up and down the buildings around you, looking for the peak from the transmitter itself.

Practicing Your Skills

It only takes two to T-hunt. You and a friend can take turns hiding and seeking. If no one else is available, try hunting local repeaters or their users. Hunting a user is particularly challenging because the signal isn't constant. In fact, it may leave the air at any minute. Of course, you'll ask permission before going onto private property. Be sure you have your amateur license with you, in case you have to prove you're not a crook plotting a heist! When you get some experience and are ready for a real challenge, try hunting mobiles!

Practical Hunting

Using FM repeaters is not without its aggravations. Microphone push-to-talk buttons sometimes stick closed. Some hams leave their rigs turned on when they leave their cars. The combination may result in unintentional jamming of a repeater. Also, Amateur Radio has its own collection of loose nuts, who sometimes feel compelled to jam repeaters or simplex frequencies. Repeater clubs often have interference committees. Committee members are probably among those who find the hidden transmitter first at the monthly T-hunt. Otherwise, they go about their committee business in secret. Only those unfortunate enough to receive a visit from the committee know how efficiently they can track down an "open mike" or a malicious jammer.

Transmitter hunting is a necessary part of high-altitude ATV (Amateur TV) balloon flights. Depending on the winds aloft, the balloon can drift as much as a hundred miles or more before it finally bursts. From the moment of launch, the hunters track the balloon, trying to position themselves directly beneath it. When the balloon pops and the payload falls earthward on its parachute, the hunters have to be ready to track the package and recover it. Thanks to the skills of transmitter hunters, most payloads are recovered successfully.

Transmitter hunting is also a skill practiced by rescue organizations, especially in the Western mountains. Many private airplanes carry emergency locator transmitters (ELTs), operating

Beam antennas are most sensitive for T-hunting, while Doppler arrays are fast and automatic. Bill (KC7VV) and Boni (WT7Z) Pineups use both. ***(KØOV photo)***

a few megahertz below the 2-meter band. ELTs emit a distinctive tone, and have built-in batteries that will keep them going for several hours. Many search-and-rescue volunteers are hams.

On With the Hunt!

The best way to find out about T-hunting in your area is to ask other users of your local repeaters. If nobody's doing it yet, why not get the ball rolling? You'll need someone to operate the hidden transmitter; this person must have at least a Technician license to operate on the 2-meter band. If you hold the hunt on the 222-MHz band, any licensed ham can be the transmitter operator.

Participants will need adequate equipment. If you don't have enough rigs to go around, form teams. Building some T-hunting antennas and attenuators would be another good club project. Imagine the pleasure you'll get from tracking down the hidden transmitter with an antenna you built yourself!

The first hunts should be easy, to give everyone a chance to find the transmitter. Arrange a meeting place and time and spread the word through club meetings and local repeaters. Depending on the terrain, you may want to distribute photocopied street maps of the area in which the transmitter may be located.

The winners of *sprint* hunts within the boundary of a town or small city are often determined by how long it takes them to find the transmitter. The team that finds the transmitter first wins. This method is common when the hunt is held on a repeater input frequency, to reduce the time the repeater is tied up. Wide-area and other advanced hunts are generally scored by mileage, much like a sports-car rally. Scoring by mileage recognizes skill in taking bearings, and discourages high-speed driving during the hunt. These hunts are usually held on simplex frequencies to avoid tying up a repeater.

If you wish, you can collect a small entry fee from each hunter, and use the proceeds to buy a plaque or an ARRL book for the winner. The best prize might be to let the winner be the transmitter operator next time. If the winner can't operate a transmitter on the band you use for your hunts, find someone to go along as control operator.

One chapter can't do justice to T-hunting; it would take a big, thorough book. For further reading—and a list of direction-finding equipment manufacturers—see the Resources and References Guide. If you're looking for an exciting Amateur Radio activity, transmitter hunting is an excellent choice. It's hard to top the thrill of chasing down a fox and savoring the satisfaction of finally finding it. Try it yourself and see!

CHAPTER 9

There is Life Above VHF!

By Rus Healy, NJ2L

What do the UHF-and-higher bands have to offer? For one thing: *room*! Sure, the VHF bands are big, but the spectrum available above 450 MHz makes VHF look like a small garden plot. The world above VHF beckons the technically inclined ham in the same way the American frontier beckoned the pioneers. When you have so much spectrum at your disposal, anything is possible!

The UHF and microwave bands don't often provide the spectacular propagation occasionally found on VHF, but you'll find plenty of surprises just the same. For example, if you thought that microwave transmissions were strictly line-of-sight, think again. Under the proper conditions, microwave contacts have been made over distances of hundreds—and even thousands—of miles!

In earlier times, UHF and microwave equipment had a reputation for being difficult to build and use. Not anymore! The availability of inexpensive no-tune transverter kits has revolutionized amateur microwave activity. No-tune means exactly what it says. You simply built the kit (or buy the product), plug in the proper cables and you're on the air!

What You Can Do . . . and Where Can You Do It

US Amateur Radio UHF allocations above 450 MHz include bands at 902-928 MHz, 1240-1300 MHz and 2300-2310/2390-2450 MHz. The next higher realm is SHF (for *super high frequency*), in which we have bands at 3300-3500 MHz, 5650-5850 MHz, 10.0-10.5 GHz and 24.0-24.25 GHz. Most amateurs refer to all the ham bands from 902 MHz (33 cm) and up as microwave bands. However, the technical definition of *microwave* is any wavelength shorter than 30 cm—which corresponds to frequencies above 1000 MHz (1 GHz). We'll go along with the common usage and refer to all the bands from 33 cm and up as microwave bands.

Activity on the microwave bands, like the VHF bands, centers around calling frequencies. Considering the size of most UHF and microwave bands, searching for individual signals would be a tedious process. Instead, operators meet on designated calling frequencies. When everyone is listening for contacts on the same frequency, it's easy to get a QSO started. Calling-frequency etiquette dictates that once you make contact, you and your partner should

Fig 9-1—Dale, AF1T, operates during the 1991 10-GHz contest from the summit of Mt Kearsarge in New Hampshire. When you don't own a tall tower, a hill or mountain is the perfect substitute.

slide off to another frequency to chat. This leaves the frequency clear for someone else to try their luck. Table 9-1 shows the microwave-band calling frequencies, band edges and the metric designators by which most hams refer to these bands.

The amateur microwave bands are available to US amateurs licensed at or above the Technician Class level. Novice licensees also have privileges on the 23-cm band, but with frequency and transmitter-power restrictions. Technician and higher-class licensees have full amateur privileges, including a 1.5-kW output-power limit, on these bands. As you'll see, however, most operating takes place with less than 20 W of transmitter output.

High power (more than a few hundred watts) is very difficult—and very expensive—to generate at frequencies above 450 MHz. Even record-setting QSOs of several thousand miles are easily made with low power when propagation enhancement opens paths to distant stations.

Any number of modes are available for you to use. SSB and

Table 9-1

US Amateur Radio Microwave Bands

Band Edges (MHz)	*Metric Designator*	*Calling Frequency (MHz)*
902-928*	33 cm	903.1
1240-1300*	23 cm	1296.1
2300-2310, 2390-2450	13 cm	2304.1
3300-3500	9 cm	3456.1
5650-5925	5 cm	5760.1
10,000-10,500	3 cm	10368.1

*Geographical and power limitations apply to some US amateurs on these bands; see *The FCC Rule Book*, available from ARRL, for details.

UHF/Microwave Equipment and Antennas

Unless you operate on the 10- or 24-GHz bands (where commercial transceivers are readily available), you'll need *transverters* to work the rest of the bands above 23 cm. Transverters get their name from their functions; they incorporate *trans*mit and receive con*verters* that use a common intermediate frequency (usually 144 or 28 MHz). This means you can use an HF or VHF multimode radio as your basic rig and add a transverter for each band you'd like to operate.

Commercial Versus Home-Brew

Building UHF and microwave ham gear used to be the province of mysterious technical wizards with access to lots of expensive, sophisticated test equipment. The wizards are still around, but many new microwave operators are getting on the air with simple, easy-to-assemble equipment.

Fig A—This no-tune transverter (left) for 1296 MHz costs about $150 to build, puts out about 20 milliwatts and has a very sensitive receiver. It requires a few milliwatts of drive from a 144-MHz transceiver, an antenna, and a simple 576-MHz local oscillator (right). Adequate for local communications, this transverter needs only a simple power amplifier to be useful for long-distance communications.

Not only that, it's now genuinely *cheap* to get on these bands! If you already have a 2-meter multimode radio, you can get a low-power signal on any band from 902 through 3456 MHz for under $200 (including antenna). We owe this ease and economy to advances made principally by Rick Campbell, KK7B, and Jim Davey, WA8NLC, who have developed no-tune transverter systems for the 902, 1296, 2304, 3456 and 5760-MHz bands. These designs use band-pass filters printed on PC boards—and they require no adjustments! Anyone with an FM-broadcast receiver can verify their operation by tuning in the local-oscillator signal (in the 90- to 95-MHz range). Fig A shows a basic single-board, no-tune 1296-MHz transverter designed by KK7B along with its companion local oscillator.

Preamplifiers and Power Amplifiers

Like microwave transverters, no-tune microwave designs have made building your own preamplifiers simple and inexpensive. Of course, commercial preamplifiers are also available, and the cost gulf between commercial and home-brew preamplifiers isn't as great as the one between corresponding transverters.

"High power" has a completely different meaning on the microwaves than on the lower-frequency bands. For the most part, if you're running more than 25 watts on any band above 1296 MHz, you're using unusually high power. With commonly used microwave antennas, you can work wonders with a few watts. Many of the current terrestrial distance records were set using low to moderate power and relatively small antennas (one or two loop Yagis and small dishes with similar gains).

Recently, however, high-gain, linear *hybrid* amplifier modules have become available. These modules make building a power amplifier easier than ever before. It's simply a matter of supplying 50-ohm input and output connections, a heat sink, power-supply and bias voltages. For about $75, you can build a 15-watt, 1296-MHz power amplifier that requires half a watt of drive and a few amperes at 13.5 V dc.

Antennas

For the 903- and 1296-MHz bands, most hams use *loop*

Continued on page 9-6

Fig B—An array of loop Yagis makes a small, light and very effective antenna system for the microwave bands. This system consists of a pair of 12-foot-boom 1296-MHz "loopers" in the background and a pair of 6¾-foot-boom 2304-MHz loopers in front of it. The whole system is easily put up by one person and turned by a light-duty TV-antenna rotator. This antenna system, used during a recent ARRL June VHF QSO Party, yielded contacts of more than 200 miles on 2304 MHz when fed with only 9 watts of transmitter power, and over 300 miles on 1296 MHz when driven with about 100 watts. At 2304, a pair of loop Yagis like this is roughly equivalent to a 30-inch dish.

Yagis, which were popularized in England and quickly spread around the world. Fig B shows four loop-Yagi antennas. "Loopers," as they're sometimes called, are easy to build and handle, have no sharp edges and resist damage well. Using tin shears, sheet aluminum and a small-diameter aluminum boom, you can build your own loop Yagis from designs in *The ARRL Antenna Book*

It doesn't get much easier than this. A hybrid amplifier module and a few capacitors provides 15 watts output with only ½ watt input on 1296 MHz.

and *The ARRL Handbook*. Inexpensive kits and ready-made loop Yagis are also available from Down East Microwave and other manufacturers.

The Gray Zone

At and above 2304 MHz, dishes become more practical than loop Yagis. For instance, a 45-element, 2304-MHz loop Yagi on a 6-3/4-foot boom requires a significant amount of labor to produce. On the other hand, a 2-foot dish has about the same gain, costs about the same and is considerably easier to transport than a long antenna. And, unlike loop Yagis, dishes can be used on several bands. Here again, no-tune dish feed designs have made it easier to use dishes. One feed designed by Tom Hill, WA3RMX, even covers three bands! (See Fig C.)

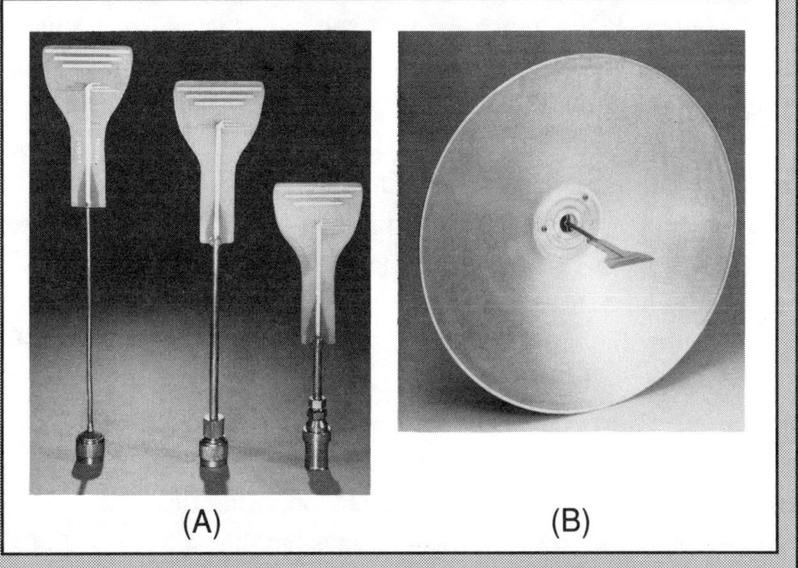

Fig C—At A, three versions of a no-tune dish feed, designed by Tom Hill, WA3RMX. This feed works on the 2304, 3456 and 5760-MHz bands. It's etched on common G10 glass-epoxy PC board material and costs very little (under $20) to make. B shows the feed mounted in a 19-inch dish.

CW are popular when you're trying to go for maximum distance. If you're more concerned about audio quality, FM is also available. Amateur television (ATV) is becoming popular in the 902-MHz band and packet is showing up just about everywhere! The sheer size of the UHF and microwave bands allows ample room for everyone to use their favorite modes with little concern for interference.

Propagation—Getting From Here to There

All forms of ionospheric propagation disappear above the 420-MHz band, so microwave propagation is limited to tropospheric phenomena and reflections of various kinds. (The troposphere is the portion of our atmosphere that reaches 7 to 10 miles above the surface of the earth. It's the region where most of our weather takes place.) Tropospheric *ducts* can produce tremendous long-distance communications opportunities. This type of propagation is responsible for most of the distance records on the bands from 902 MHz and up. Tropospherically ducted signals are usually very strong. One legendary 2500+-mile duct opens regularly between California and Hawaii. It has allowed N6CA and KH6HME to work each other on all the microwave bands through 5760 MHz!

During "flat" times (when there's no tropospheric enhancement), over-the-horizon microwave radio paths are generally supported by tropospheric scatter. Scatter signals are usually weak, and often undergo considerable fading. Even so, they provide some exciting microwave DX opportunities.

All other microwave propagation is facilitated by reflections. Any metallic object that's at least 10 wavelengths across can reflect microwave RF. Signs, water towers, airplanes and many other objects fall into this category. For example, you may not be able to communicate with a friend because there's a large hill blocking your microwave signals. Is there a water tower or some other sizable metal object that both of you can see from your respective houses? Try bouncing your signals off its surface and you may be surprised to discover that you can hear each other just fine! This is a form of microwave billiards that commercial users frequently employ. The next time you're traveling through a mountainous region, see if you can spot what appear to be blank billboards perched on ridges and hillsides. Chances are these are microwave reflectors providing communication paths to hard-to-reach areas.

When it comes to reflections, even the moon gets into the act: Amateurs have made moonbounce (EME) contacts on all amateur bands from 50 MHz through 10 GHz! The most popular microwave EME band is 1296 MHz, with 2304 MHz a relatively distant second. Few moonbounce-capable stations exist for the 903, 3456, 5760 and 10,368-MHz bands, but many stations have made EME contacts on these bands. Most EME contacts take place through prearranged schedules. ("I'll begin transmitting on 1296.010 MHz at 0300 UTC. You and I should both be able to see the moon at that time.")

Where the Action Is

Like most VHF activity, you'll find the majority of microwave stations near population centers, particularly in the Northeast. Large groups of microwave-capable stations exist near Philadelphia and in surrounding areas; the Rochester, New York, area; and Connecticut and eastern Massachusetts. You don't need to live in one of these areas to make microwave operation worthwhile, though. These bands are substantially populated in the Central US, particularly northern Texas and the plains states, and near the California coast between San Francisco and San Diego. Other areas are booming in microwave growth.

You can get a good idea whether there's microwave activity in your area by being active on the 144- and 420-MHz bands, which many stations use to coordinate microwave contacts.

Because of the relatively small number of hams on the microwave bands, you generally won't find operators CQing except on activity nights, and during contests and band openings (see Chapter 4). Like moonbounce contacts, most microwave QSOs are scheduled beforehand. Again, familiarity with local VHF activity is your best guide to what's happening on the microwaves.

Get On the Air and GO!

The only limitations in microwave communications are the ones imposed by your imagination. Start collecting your equipment and making plans to get on the air right away. Visit a hamfest flea market and look for used parabolic dishes. They're cheap and they'll get you started toward creating a very powerful

Fig 9-2—When it comes to contests, many clubs get into the act in a big way! The members of the San Bernardino Microwave Society set up camp on Heaps Peak in southern California. With their 4-foot and 30-inch parabolic dishes, they made an impressive number of contacts and had a good time, too!

antenna system. Check out the loop Yagis, too.

In the Resources and References Guide you'll find books and articles that will help you assemble your first UHF and microwave station. You say you don't have a 100-foot tower for your antennas? No problem. During the next contest or activity night, take your equipment to the top of a hill, a mountain or even a tall building. Bring along your Amateur Radio friends and make a party of it!

How about setting up a permanent microwave link with a friend? You can link your computers with packet radio, exchange ATV video or just chat. Depending on where you and your friend live, small roof-mounted antennas may be adequate.

And don't forget the satellites! Many Amateur Radio satellites have microwave capability. See the satellite frequency tables in Chapter 5.

If you use your imagination to explore our rich microwave resources, you may stumble onto something innovative and fascinating. For example, some amateurs have experimented with using microwaves to create their own weather radar systems. They aim their antennas at approaching storms and probe them with pulsating signals. By using computers to interpret the echoes, they can gauge their distance, speed and many other characteristics.

Reading about UHF and microwaves is interesting, but the real fun begins when you close the books and start *operating*. If nothing else, find microwave enthusiasts in your area and ask to join them during their next contest operation. If that doesn't infect you with the microwave "bug," nothing will!

Resources and References Guide

Unless otherwise indicated, all items in this section may be purchased, subject to availability, from your dealer or from the American Radio Relay League, 225 Main St, Newington, CT 06111; tel 203-666-1541, fax 203-665-7531. Prices subject to change without notice.

FM and Repeaters

Books

The ARRL Handbook for Radio Amateurs—Covers analog and digital electronics theory, antennas, transmitters, receivers, processing equipment and accessories. Chapter 14 discusses the technical aspects of FM repeaters.

The ARRL Operating Manual—The most complete guide to Amateur Radio operating ever published. Chapter 11, by Brian Battles, WS1O, is devoted to FM and repeater operating. ARRL Order No. 1086, $18.

The ARRL Repeater Directory— The annually published pocket-sized listing of 20,000 repeaters in the US and Canada on 29, 50-54 and 144-148, 222-225, 420-450, 902-928 and 1240 MHz and above, including FM voice, packet radio and amateur television repeaters. Operating tips, band plan charts, club listings and more. ARRL Order No. 3797, $6.

Packet Radio

Books

Your Gateway to Packet Radio by Stan Horzepa, WA1LOU. A thorough discussion of packet radio, including many of the latest developments. ARRL Order No. 2030, $12

The ARRL Operating Manual, 4th edition. In Chapter 10, Stan Horzepa, WA1LOU, describes how to set up a packet station. He also addresses operating techniques and various networking systems such as NET/ROM, TCP/IP, ROSE and TEXNET. ARRL Order No. 1086, $18

AX.25 Amateur Packet-Radio Link-Layer Protocol—Ideal for the ham who wants to learn the inner workings of the AX.25 packet protocol. ARRL Order No. 0119, $8

Newsletters

Packet Status Register—Published quarterly by Tucson Amateur Packet Radio (TAPR), PO Box 12925, Tucson, AZ 85732. $15/year.

Packet Satellite Software

AMSAT provides software for accessing the PACSATs and decoding telemetry from a variety of satellites. For a complete list of available software, send a self-addressed, stamped envelope to AMSAT, PO Box 27, Washington, DC 20044; tel 301-589-6062.

SSB and CW

Books

The ARRL Operating Manual—Michael Owen, W9IP, takes a

fascinating look at VHF/UHF operating in Chapter 12. ARRL Order No. 1086, $18

Radio Auroras—Charlie Newton, G2FKZ, details the interesting and unpredictable world of communication via auroral propagation in this Radio Society of Great Britain book. ARRL Order No. 3568, $18

VHF/UHF Manual—Published by the Radio Society of Great Britain. Includes information on the history of VHF/UHF communications, propagation, receivers, transmitters, antennas and much more. ARRL Order No. R630, $30

ARRL World Grid Locator Atlas, ARRL Order No. 2944, $5

Maps

ARRL Grid Locator Map, ARRL Order No. 1290, $1

Newsletters

50 MHz DX Bulletin—Shel Remington, NI6E/KH6, PO Box 1222, Keaau, HI 96749. 18 issues/year. $20.

220 Notes—A. B. Reis, K9XI, W6539 Birch St, Onalaska, WI 54650-9001. Send a self-addressed, stamped envelope for more information.

432 and Above EME News—Allen Katz, K2UYH, Electronic Engineering Dept, Trenton State College, Trenton, NJ 08650-4700.

Midwest VHF Report—Art Hambleton, K1LLO, 10004 High Meadows, Blackhawk, SD 57718; tel 605-787-4552. 9 issues/year. Send a self-addressed, stamped envelope for more information.

Terrestrial VHF—John Carter, KØIFL, PO Box 554, Union, MO 63084. Send a self-addressed, stamped envelope for more information.

The West Coast VHFer—Ham Tech, PO Box 685, Holbrook, AZ 86025; tel 602-524-3354. 12 issues/year. $14

Organizations

Six Meter International Radio Klub (SMIRK), c/o Ray Clark, K5ZMS, 7158 Stone Fence Dr, San Antonio, TX 78227. Dues are $6 per year. To apply for membership you must confirm contact with at least six SMIRK members. (Submit a log sheet indicating the date, time, frequency, call sign and membership number of each contact.)—*information provided by Joe Lynch, N6CL*

Side Winders on Two (SWOT), c/o Howard Hallman, WD5DGJ, 3230 Springfield, Lancaster, TX 75134. Dues are $10 per year. To apply for membership you must confirm contact with at least two SWOT members. (Submit a log sheet indicating the date, time, frequency, call sign and membership number of each contact.)—*information provided by Joe Lynch, N6CL*

Software

GRIDLOC—A program for IBM PCs and compatibles that determines the grid square designations for various locations. It is one of many useful programs on *The ARRL UHF/Microwave Experimenter's Manual* companion disk (5.25-inch). ARRL Order No. 3134, $10.

Satellite Communications

Books

The ARRL Operating Manual—In Chapter 13, Jon Bloom, KE3Z, offers a comprehensive—yet easy to understand—look at amateur satellite operating. ARRL Order No. 1086, $18.

Satellite Experimenter's Handbook—Martin Davidoff, K2UBC, provides the ultimate reference for the satellite operator. All active satellites are covered in detail, including telemetry formats, uplink/downlink frequencies, on-board power systems and more. ARRL Order No. 3185, $20.

ARRL Satellite Anthology—A collection of the best satellite articles from recent issues of *QST*. A must for every satellite operator. ARRL Order No. 3800, $8.

Decoding Telemetry from Amateur Satellites—G. Gould Smith, WA4SXM, shows you how to receive and decode telemetry signals. Keep your fingers on the pulse of our active amateur satellites! Available from AMSAT, PO Box 27, Washington, DC 20044; tel 301-589-6062. $15

Newsletters

The AMSAT Journal—available from AMSAT, PO Box 27, Washington, DC 20044; tel 301-589-6062. $30/year.

OSCAR Satellite Report—available from R. Myers Communications, PO Box 17108, Fountain Hill, AZ 85269-71108. $56/year US, $62/year Canada.

Software

Software for satellite tracking, telemetry decoding and Pacsat operation is available from: AMSAT, PO Box 27, Washington, DC 20044; tel 301-589-6062. Send a self-addressed, stamped envelope and ask for their software catalog.

PSK/FSK Pacsat Modems

Tucson Amateur Packet Radio, PO Box 12925, Tucson, AZ 85732; tel 602-749-9479.

PacComm Inc, 4413 N Hesperides St, Tampa, FL 33614-7618; tel 813-874-2980.

Orbital Elements

Orbital elements for all active Amateur Radio satellites are transmitted twice weekly by W1AW. See the W1AW schedule in this section.

Orbital elements are also available on AMSAT Information Nets:

International: Sunday at 1800 UTC on 21.280 and 14.282 MHz
International: Saturday at 2200 UTC on 21.280 MHz
US East Coast: Tuesday at 2100 EST/EDT on 3.840 MHz
US Central: Tuesday at 2100 CST/CDT on 3.840 MHz
US West Coast: Tuesday at 2100 PST/PDT on 3.840 MHz

ATV

Periodicals

Amateur Television Quarterly—1545 Lee St, Suite 73, Des Plaines, IL 60018, tel 708-298-2269. $18/year US, $22/year Canada

The SPEC-COM Journal—PO Box 1002, Dubuque, IA 52004; tel 319-557-8791. $20/year US, $25/year Canada

Equipment

Advanced Electronic Applications Inc, PO Box C2160, 2006 196th St SW, Lynnwood, WA 98036-0918; tel 206-775-7373, fax 206-775-2340.

Communication Concepts Inc, 508 Millstone Dr, Xenia, OH 45385; tel 513-426-8600.

Down East Microwave, Box 2310 RR1, Troy, ME 04987; tel 207-948-3741

Elktronics, 12536 Township Road 77, Findlay, OH 45840; tel 419-422-8206.

Micro Computer Concepts, 7869 Rustic Woods Dr, Dayton, OH 45424; tel 513-233-9675.

Micro Video Products, 1334 S Shawnee Dr, Santa Ana, CA 92704; tel 800-473-0538 or 714-957-9268.

PC Electronics, 2522 Paxson Ln, Arcadia, CA 91007-8537; tel 818-447-4565, fax 818-447-0489.

Supercircuits, 1403 Bayview Dr, Hermosa Beach, CA 90254; tel 310-372-9166.

Tactical Electronics Corp, PO Box 1743, Melbourne, FL 32902; tel 407-676-6907, fax 407-951-4630.

Wyman Research Inc, RR #1 Box 95, Waldron, IN 46182; tel 317-525-6452.

Computer Bulletin Boards

On CompuServe, check the HamNET forums (enter GO HAMNET).

The Electronic Cottage BBS is run by the US ATV Society (USATVS) and Spec-Com at 319-582-3235, SysOps Bill Fay, KAØFDI; Mike Stone, WBØQCD; Jim Bussan and Pat Powers. The Electronic Cottage also has an exhaustive listing of other ham-oriented BBSs nationwide.

Another ATV-oriented BBS is operated by Howard Bacon, KC4CIQ, of South Pittsburg, Tennessee, at 615-837-8352.

Awards

Books

CQ USA County Award Record Book—$1.25 from CQ Publishing, 76 N Broadway, Hicksville, NY 11801.

The ARRL Operating Manual—Chapter 8 is devoted entirely to awards, with complete details concerning many VHF award programs as well as color reproductions of award certificates. Chapter 17 contains an official counties list. ARRL Order No. 1086, $18.

Maps

ARRL Grid Locator Map, ARRL Order No. 1290, $1

Organizations

Six Meter International Radio Klub (SMIRK), c/o Ray Clark, K5ZMS, 7158 Stone Fence Dr, San Antonio, TX 78227. Dues are $6 per year. To apply for membership you must confirm contact with at least six SMIRK members. (Submit a log sheet indicating the date, time, frequency, call sign and membership number of each contact.)—*information provided by Joe Lynch, N6CL*

Side Winders on Two (SWOT), c/o Howard Hallman, WD5DGJ, 3230 Springfield, Lancaster, TX 75134. Dues are $10 per year. To apply for membership you must confirm contact with at least two SWOT members. (Submit a log sheet indicating the date, time, frequency, call sign and membership number of each contact.)—*information provided by Joe Lynch, N6CL*

Transmitter Hunting

Equipment

BMG Engineering, 9935 Garibaldi, Temple City, CA 91780

Doppler Systems, PO Box 2780, Carefree, AZ 85377; tel 602-488-9755

L-Tronics, 5546 Cathedral Oaks Rd, Santa Barbara, CA 93111; tel 805-967-4859

Radio Engineers, 3941 Mt Brundage, San Diego, CA 92111; tel 619-565-1319

Books

Transmitter Hunting—by Joseph Moell, KØOV, and Thomas Curlee, WB6UZZ. ARRL Order No. 2701, $19.

UHF and Microwave

Books

The ARRL UHF/Microwave Experimenter's Manual—All

amateurs will find this book useful as the basis for understanding microwave technology. The book includes information on design and fabrication techniques, propagation, antennas and much more. ARRL Order No. 3126, $20

Microwave Handbook, Volume 1—From the Radio Society of Great Britain, the Handbook covers operating techniques, systems analysis, antennas, microwave semiconductors and tubes. ARRL Order No. 2901, $35.

Microwave Handbook, Volume 2—Construction tips, equipment design, microwave beacons and repeaters, test equipment and more. ARRL Order No. 3606, $35.

Articles

R. Campbell, KK7B, "Getting Started on the Microwave Bands," *QST*, Feb 1992, pp 35-39.

E. Pocock, W3EP, "Getting from Here to There on 2304-MHz," *QST*, Nov 1988, pp 15-16.

A. Ward, WB5LUA, "Simple Low-Noise Microwave Preamplifiers," *QST*, May 1989, pp 31-36, 75.

Z. Lau, KH6CP, "SHF Systems SHF 1240K 1296-MHz Transverter Kit," Product Review, *QST*, Apr 1990, pp 33-34.

D. Mascaro, WA3JUF, "A High-Performance UHF and Microwave System Primer," *QST*, May 1991, pp 30-33.

R. Campbell, KK7B, "A Single-Board, No-Tune 902-MHz Transverter," *QST*, Jul 1991, pp 25-29; Feedback, *QST*, Feb 1992, p 39.

J. Davey, WA8NLC, "A No-Tune Transverter for 3456 MHz," *QST*, Jun 1989, pp 21-26; Feedback, *QST*, Oct 1990, p 31.

R. Campbell, KK7B, "A Single-Board Bilateral 5760-MHz

Transverter," *QST*, Oct 1990, pp 27-31.

T. Hill, WA3RMX, "A Triband Microwave Dish Feed," *QST*, Aug 1990, pp 23-27. Describes an inexpensive PC-board dish feed for 2304, 3456 and 5760 MHz.

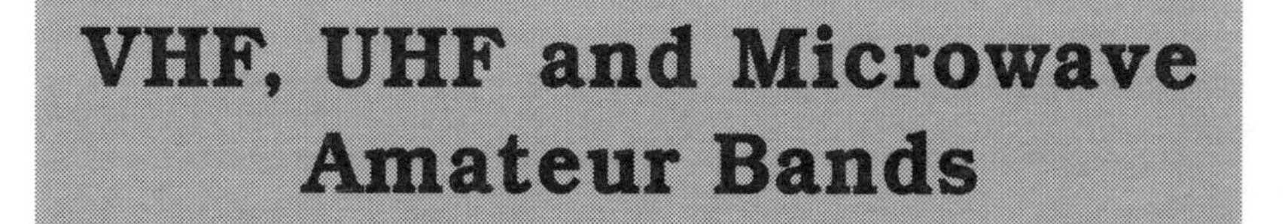

VHF, UHF and Microwave Amateur Bands

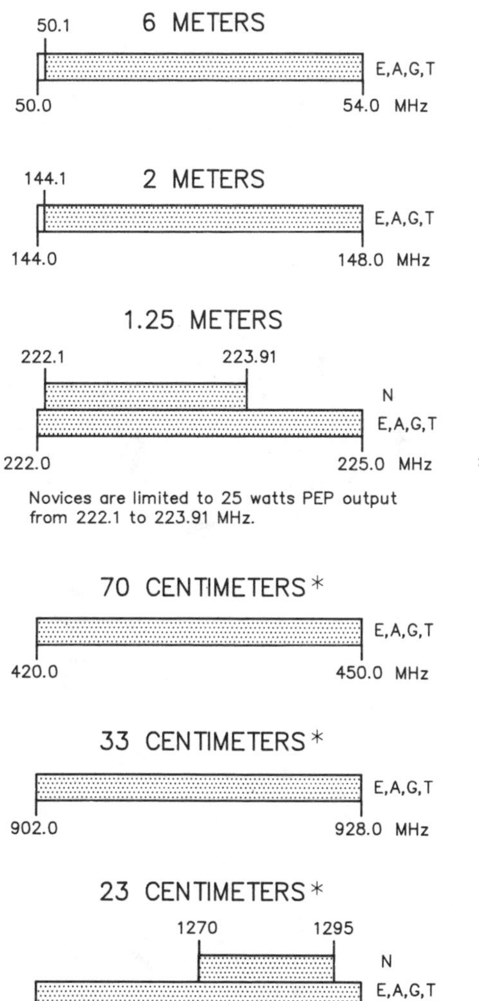

Novices are limited to 25 watts PEP output from 222.1 to 223.91 MHz.

Novices are limited to 5 watts PEP output from 1270 to 1295 MHz.

KEY

[dotted box] = CW, RTTY, data, MCW, test, phone and image

[empty box] = CW only

E = AMATEUR EXTRA
A = ADVANCED
G = GENERAL
T = TECHNICIAN
N = NOVICE

* Geographical and power restrictions apply to these bands See The FCC Rule Book for more information about your area.

Above 23 Centimeters:

All licensees except Novices are authorized all modes on the following frequencies:
2300–2310 MHz
2390–2450 MHz
3300–3500 MHz
5650–5925 MHz
10.0–10.5 GHz
24.0–24.25 GHz
47.0–47.2 GHz
75.5–81.0 GHz
119.98–120.2 GHz
142–149 GHz
241–250 GHz
All above 300 GHz

ARRL 50, 144, 222 and 420-MHz Band Plans

MHz

- 50.00-50.10 CW/Beacons
  - 50.06-50.08 Automatically Controlled beacons
- 50.10-50.30 SSB, CW
  - 50.10-50.125 DX Window
  - 50.125 SSB Calling
- 50.30-50.6 All Modes
- 50.60-50.80 Nonvoice Communications
  - 50.62 Digital calling
- 50.80-51.00 Radio Control (20 kHz channels)
- 51.00-51.10 Pacific DX Window
- 51.12-51.48 Repeater inputs (19 channels)
  - 51.12-51.18 Digital repeater inputs
- 51.50-51.60 Simplex (6 channels)
- 51.62-51.98 Repeater outputs (19 channels)
  - 51.62-51.68 Digital repeater outputs
- 52.0-52.48 Repeater inputs (except as noted; 23 channels)
  - 52.02, 52.04 FM simplex
  - 52.2 TEST PAIR (input)
- 52.50-52.98 Repeater outputs (except as noted; 23 channels)
  - 52.525 Primary FM simplex
  - 52.54 Secondary FM simplex
  - 52.7 TEST PAIR (output)
- 53.0-53.48 Repeater inputs (except as noted; 19 channels)
  - 53.0 Remote base FM simplex
  - 53.02 Simplex
  - 53.1, 53.2, 53.3, 53.4 Radio remote control
- 53.50-53.98 Repeater outputs (except as noted; 19 channels)
  - 53.5, 53.6, 53.7, 53.8 Radio remote control
- 53.52, 53.9 Simplex

MHz

- 144.00-144.05 EME (CW)
- 144.05-144.06 General CW and weak signals
- 144.06-144.10 General CW and weak signals
- 144.10-144.20 EME and weak-signal SSB
  - 144.20 National calling
- 144.20-144.275 General SSB operation
- 144.275-144.30 Beacons
- 144.30-144.50 New OSCAR subband
- 144.50-144.60 Linear translation inputs
- 144.60-144.90 FM repeater inputs
- 144.90-145.10 Weak signal and FM simplex
- 145.10-145.20 Linear translator outputs
- 145.10-145.50 FM repeater outputs
- 145.50-145.80 Miscellaneous and experimental modes
- 145.80-146.00 OSCAR subband
- 146.01-146.37 Repeater inputs
- 146.40-146.58 Simplex
- 146.61-147.39 Repeater outputs
- 147.42-147.57 Simplex
- 147.60-147.99 Repeater inputs

MHz

- 222.0-222.15 Weak-signal modes
  - 222.0-222.025 EME
  - 222.05-222.06 Propagation beacons
  - 222.10 Calling frequency (SSB & CW)
  - 222.10-222.15 Weak-signal SSB & CW
- 222.15-222.25 General operation, CW or SSB, etc.
- 222.25-223.38 Repeater inputs
- 223.40-223.52 Simplex
- 223.52-223.64 Digital, packet
- 223.64-223.70 Links, control
- 223.71, 223.85 Simplex, packet, repeater outputs
- 223.85-224.98 Repeater outputs

MHz

- 420.00-426.00 ATV repeater or simplex with 421.25-MHz video carrier, control links and experimental
- 426.00-432.00 ATV simplex with 427.25-MHz video carrier
- 432.00-432.07 EME
- 432.07-432.10 Weak-signal CW
- 432.10 Calling frequency
- 432.10-432.30 Mixed mode and weak signal
- 432.30-432.40 Beacons
- 432.40-433.00 Mixed mode and weak signals
- 433.00-435.00 Auxiliary/repeater links
- 435.00-438.00 Satellite only (internationally)
- 438.00-444.00 ATV repeater input with 439.25-MHz video carrier and repeater links
  - 442.00-445 Repeater inputs and outputs (local option)
- 445.00-447.00 Shared by auxiliary and control links, repeaters and simplex (local option)
  - 446.00 National simplex frequency
- 447.00-450.00 Repeater inputs and outputs (local option)

W1AW Schedule

MTWThFSSn = Days of Week   Dy = Daily

W1AW code practice and bulletin transmissions are sent on the following schedule:

ET	Slow Code Practice	WF: 9 AM; MWF: 7 PM; TThSSn: 4 PM, 10 PM
	Fast Code Practice	MWF: 4 PM, 10 PM; TTh: 9 AM; TThSSn: 7 PM
	CW Bulletins	Dy: 5 PM, 8 PM, 11 PM; TWThF: 10 AM
	Teleprinter Bulletins	Dy: 6 PM, 9 PM, 12 PM; TWThF: 11 AM
	Voice Bulletins	Dy: 9:45 PM, 12:45 AM
CT	Slow Code Practice	WF: 8 AM; MWF: 6 PM; TTHSSn: 3 PM, 9 PM
	Fast Code Practice	MWF: 3 PM, 9 PM; TTh: 8 AM; TThSSn: 6 PM
	CW Bulletins	Dy: 4 PM, 7 PM, 10 PM; TWThF: 9 AM
	Teleprinter Bulletins	Dy: 5 PM, 8 PM, 11 PM; TWThF: 10 AM
	Voice Bulletins	Dy: 8:45 PM, 11:45 PM
MT	Slow Code Practice	WF: 7 AM; MWF: 5 PM; TThSSn: 2 PM; 8 PM
	Fast Code Practice	MWF: 2 PM, 8 PM; TTh: 7 AM; TThSSn: 5 PM
	CW Bulletins	Dy: 3 PM, 6 PM, 9 PM; TWThF: 8 AM
	Teleprinter Bulletins	Dy: 4 PM, 7 PM, 10 PM; TWThF: 9 AM
	Voice Bulletins	Dy: 7:45 PM, 10:45 PM
PT	Slow Code Practice	WF: 6 AM; MWF: 4 PM; TThSSn: 1 PM; 7 PM
	Fast Code Practice	MWF: 1 PM, 7 PM; TTh: 6 AM; TThSSn: 4 PM
	CW Bulletins	Dy: 2 PM, 5 PM, 8 PM; TWThF: 7 AM
	Teleprinter Bulletins	Dy: 3 PM, 6 PM, 9 PM; TWThF: 8 AM
	Voice Bulletins	Dy: 6:45 PM, 9:45 PM

Code Practice, Qualifying Run and CW bulletin frequencies: 1:818, 3.5815, 7.0475, 14.0475, 18.0975, 21.0675, 28. 0675, 147.555 MHz.

Teleprinter bulletin frequencies: 3.625, 7.095, 14.095, 18.1025, 21.095, 28.095, 147.555 MHz.
Voice bulletin frequencies: 1.89, 3.99, 7.29, 14.29, 18.16, 21.39, 28.59, 147.555 MHz

Slow code practice is at 5, 7 ½, 10, 13 and 15 WPM. Fast code practice is at 35, 30, 25, 20, 15, 13 and 10 WPM.

Code practice texts are from *QST,* and the source of each practice is given at the beginning of each practice and at the beginning of alternate speeds. For example, "Text is from September 1990 *QST,* pages 16 and 79" indicates that the main text is from the article on page 16 and the mixed number/letter groups at the end of each speed are from the contest sources on page 79.

Some of the slow practice sessions are sent with each line of text from *QST* reversed. For example, "Last October, the ARRL Board of Directors" would be sent as DIRECTORS OF BOARD ARRL, OCTOBER LAST.

Teleprinter bulletins are 45.45-baud Baudot and 100-baud AMTOR, FEC mode B. ASCII (110-baud) is sent as time allows.

On Tuesdays and Saturdays at 2330 UTC, orbital elements for active amateur satellites will be sent on the regular teleprinter frequencies.

CW bulletins are sent at 18 WPM.

W1AW is open for visitors Monday through Friday from 9 AM to 11 PM Eastern time and on Saturday and Sunday from 4:30 PM to 11 PM Eastern time. If you desire to operate W1AW, be sure to bring a copy of your license with you. W1AW is available for operation by visitors between 1 and 4 PM Monday through Friday.

In a communications emergency, monitor W1AW for special bulletins as follows: voice on the hour, teleprinter at 15 minutes past the hour, and CW on the half hour.

About The American Radio Relay League

The seed for Amateur Radio was planted in the 1890s, when Guglielmo Marconi began his experiments in wireless telegraphy. Soon he was joined by dozens, then hundreds, of others who were enthusiastic about sending and receiving messages through the air—some with a commercial interest, but others solely out of a love for this new communications medium. The United States government began licensing Amateur Radio operators in 1912.

By 1914, there were thousands of Amateur Radio operators—hams—in the United States. Hiram Percy Maxim, a leading Hartford, Connecticut, inventor and industrialist saw the need for an organization to band together this fledgling group of radio experimenters. In May 1914 he founded the American Radio Relay League (ARRL) to meet that need.

Today ARRL, with about 160,000 members, is the largest organization of radio amateurs in the United States. The League is a not-for-profit organization that:

- ❏ promotes interest in Amateur Radio communications and experimentation
- ❏ represents US radio amateurs in legislative matters, and
- ❏ maintains fraternalism and a high standard of conduct among Amateur Radio operators.

At League Headquarters in the Hartford suburb of Newington, the staff helps serve the needs of members. ARRL is also International Secretariat for the International Amateur Radio Union, which is made up of similar societies in more than 100 countries around the world.

ARRL publishes the monthly journal *QST*, as well as newsletters and many publications covering all aspects of Amateur Radio. Its Headquarters station, W1AW, transmits bulletins of interest to radio amateurs and Morse Code practice sessions. The League also coordinates an extensive field organization, which includes volunteers who provide technical information for radio amateurs and public-service activities. ARRL also represents US amateurs with the Federal Communications Commission and other government agencies in the US and abroad.

Membership in ARRL means much more than receiving *QST* each month. In addition to the services already described, ARRL offers membership services on a personal level, such as the ARRL Volunteer Examiner Coordinator Program and a QSL bureau.

Full ARRL membership (available only to licensed radio amateurs) gives you a voice in how the affairs of the organization are governed. League policy is set by a Board of Directors (one from each of 15 Divisions). Each year, half of the ARRL Board of Directors stands for election by the full members they represent. The day-to-day operation of ARRL HQ is managed by an Executive Vice President and a Chief Financial Officer.

No matter what aspect of Amateur Radio attracts you, ARRL membership is relevant and important. There would be no Amateur Radio as we know it today were it not for the ARRL. We would be happy to welcome you as a member! (An Amateur Radio license is not required for Associate Membership.) For more information about ARRL and answers to any questions you may have about Amateur Radio, write or call:

ARRL Educational Activities Dept
225 Main Street
Newington, CT 06111
(203) 666-1541

Index

(Note: "RRG" refers to the Resources and References Guide)

F

G

H

I

L

M

S

Y

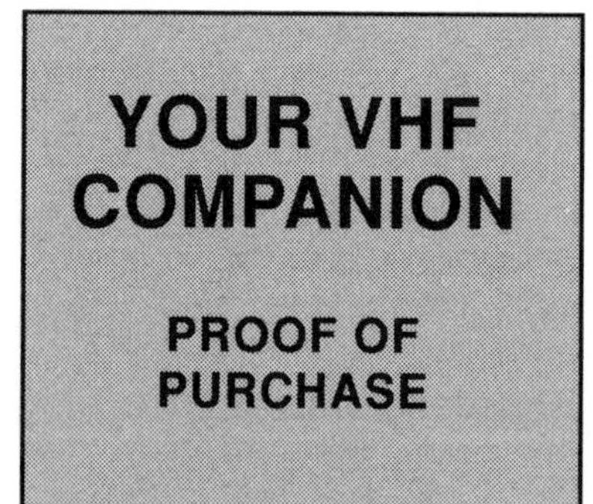
YOUR VHF
COMPANION
PROOF OF
PURCHASE

Please use this form to give us your comments on this book and what you'd like to see in future editions.

License class:

□ Novice □ Technician □ Technician with HF privileges
□ General □ Advanced □ Extra

Name ______________________ ARRL member? □ Yes □ No

Call sign ______________

Daytime Phone (   ) ______________ Age ______________

Address ____

City, State/Province, ZIP/Postal Code ______________________

If licensed, how long? ______________

Other hobbies ______________

____

Occupation ______________

For ARRL use only

Edition 1 2 3 4 5 6 7 8 9 10 11 12

Printing 1 2 3 4 5 6 7 8 9 10 11 12

From ______________________

____

____

Please affix postage. Post Office will not deliver without postage.

EDITOR, YOUR VHF COMPANION
AMERICAN RADIO RELAY LEAGUE
225 MAIN ST
NEWINGTON CT 06111

please fold and tape